AF723865

Modeling and Experimental Characterization of Nanocomposite Materials

Modeling and Experimental Characterization of Nanocomposite Materials

Special Issue Editor

Homayoun Hadavinia

MDPI • Basel • Beijing • Wuhan • Barcelona • Belgrade • Manchester • Tokyo • Cluj • Tianjin

Special Issue Editor
Homayoun Hadavinia
Kingston University
UK

Editorial Office
MDPI
St. Alban-Anlage 66
4052 Basel, Switzerland

This is a reprint of articles from the Special Issue published online in the open access journal *Nanomaterials* (ISSN 2079-4991) (available at: https://www.mdpi.com/journal/nanomaterials/special_issues/modeling_characterization_nano).

For citation purposes, cite each article independently as indicated on the article page online and as indicated below:

LastName, A.A.; LastName, B.B.; LastName, C.C. Article Title. *Journal Name* **Year**, *Article Number*, Page Range.

ISBN 978-3-03928-632-4 (Pbk)
ISBN 978-3-03928-633-1 (PDF)

© 2020 by the authors. Articles in this book are Open Access and distributed under the Creative Commons Attribution (CC BY) license, which allows users to download, copy and build upon published articles, as long as the author and publisher are properly credited, which ensures maximum dissemination and a wider impact of our publications.
The book as a whole is distributed by MDPI under the terms and conditions of the Creative Commons license CC BY-NC-ND.

Contents

About the Special Issue Editor

Homayoun Hadavinia's (Dr.) main research area is the measuring, finite element modelling, and prediction of the performance and behavior of fiber reinforced polymer (FRP) composite structures under monotonic, fatigue, and impact loadings. He has completed projects related to polymer foams, adhesively bonded joints, and lifetime prediction methodology of FRP composite and metallic structures. Currently, Dr. Hadavinia is an Associate Professor in Composite Materials at Kingston University London, working on projects related to nanocomposites for enhancement of electromechanical properties of polymer matrix using carbon nanotubes (CNTs), graphene, boron nitride nanotubes (BNNTs), boron nitride nanosheets (BNNS), and improvement of rain erosion resistant coatings for wind turbine blades.

Preface to "Modeling and Experimental Characterization of Nanocomposite Materials"

The possibility to enhance the desired properties of composites by additive nanoscale secondary reinforcement has attracted interest, both in the research arena and in industrial applications. The research in this field has been oriented toward experimental, theoretical, and numerical modelling. As a result, progress has occurred on many fronts, including manufacturing processes and scaling up; characterizing the interface between the additives and the matrix; functionalizing the nanoparticles to enhance the bonding with the matrix and the dispersion, characterizing the mechanical, thermal, and electrical properties; and theoretical/computational analysis of nanocomposite responses under different types of loading, either embedded in the fiber composites or as a polymer nanocomposites.

The papers in this Special Issue include studies of researchers from different branches of science and engineering disciplines working on experiments and modelling of nanocomposites into one volume. The seven papers presented in this volume cover experimental, computational, and theoretical aspects, dealing with many important state-of-the-art technologies and methodologies regarding the synthesis, fabrication, characterization, properties, design, and applications, and both finite element analysis and molecular dynamic simulations of nanocomposite materials and structures.

The Guest Editor is thankful to all those authors who contributed their works to this Special Issue. I hope other researchers will benefit from the results presented in the published papers in their future works.

Homayoun Hadavinia
Special Issue Editor

nanomaterials

MDPI

Article

Using the Equivalent Fiber Approach in Two-Scale Modeling of the Elastic Behavior of Carbon Nanotube/Epoxy Nanocomposite

Mahdi Javadinejad [1], Mohammad Mashayekhi [2,*], Mehdi Karevan [2] and Homayoun Hadavinia [3,*]

1 Mechanical Engineering Group, Pardis College, Isfahan University of Technology, Isfahan 84156-83111, Iran; mmjavadinejad@gmail.com

2 Department of Mechanical Engineering, Isfahan University of Technology, Isfahan 84156-83111, Iran; mkarevan@cc.iut.ac.ir

3 Department of Mechanical Engineering, Kingston University, London SW15 3DW, UK

* Correspondence: mashayekhi@cc.iut.ac.ir (M.M.); h.hadavinia@kingston.ac.uk (H.H.); Tel.: +44-20-8417-4864 (H.H.)

Received: 26 August 2018; Accepted: 4 September 2018; Published: 6 September 2018

Abstract: In this study, the mechanical behavior of epoxy/carbon nanotubes (CNTs) nanocomposite is predicated by a two-scale modeling approach. At the nanoscale, a CNT, the interface between the CNT and the matrix and a layer of the matrix around the CNT are modeled and the elastic behavior of the equivalent fiber (EF) has been identified. The CNT/epoxy interface behavior is modeled by the Park–Paulino–Roesler (PPR) potential. At the microscale, the EFs are embedded in the matrix with the extracted elastic properties from the nanoscale model. The random pattern has been used for the dispersing of EFs in the representative volume element (RVE). The effect of CNTs agglomeration in the epoxy matrix has also been investigated at the micro level. The Young's modulus of the nanocomposite was extracted from simulation of the RVE. CNT/epoxy nanocomposites at four different volume fractions were manufactured and the modeling results were validated by tensile tests. The results of the numerical models are in good agreement with the experiments and micromechanics theory, and by considering agglomeration of CNT in the model, the modeling results match with the experiments.

Keywords: carbon nanotube; interface; cohesive element; equivalent fiber; CNT agglomeration

---

1. Introduction

The use of polymer matrix composites in recent years has been growing dramatically. The recent development of nanotechnology and incorporation of nanoparticles in polymer matrix resulted in new generation of nanocomposites with enhanced properties. The result is increased use of polymer nanocomposites in various industries such as aeronautics, marine, automotive, medicine, and sports equipment [1,2]. Epoxy is a thermoset polymer with relatively light weight, low cost, high strength, and high Young's modulus which can be easily combined with other materials, in spite of limitations such as low thermal conductivity and poor flame resistance. These advantages make the epoxy an excellent choice to manufacture all kinds of polymer nanocomposites [3–6].

For establishing connections between network structure and various physical and mechanical properties such as glass transition, modulus in the glassy state and development of residual stresses series of epoxy–amine networks of well-controlled architecture were studied [7]. In addition, the yielding behavior of epoxy–amine resins was investigated by a study of the stress–strain curves in compression mode, recorded at various temperatures and strain rates. Two types of antiplasticizer were examined, depending on whether they remain miscible to the network or give rise to nanoscale phase separation along network construction, lead to improved toughness [8].

Various nanoparticles are used for improvement of mechanical properties of epoxy such as carbon nanotubes (CNTs), graphite, graphene oxide, clay, silicon nitride, silica, and nitride [9]. Among these, CNTs/epoxy nanocomposite has extremely high mechanical properties [3,10,11]. It should be noted that the use of CNTs is not limited to the strengthening of the nanocomposites. Some other properties such as low weight, high thermal stability, high electrical conductivity, and high chemical resistance make CNTs a proper candidate for enhancing nanocomposite properties [12]. Enhancing the electrical conductivity of polymers has boosted their growth in electronics and for lightning strike protection in aerospace applications.

The prediction of the mechanical properties of nanocomposites by numerical methods is a very suitable solution for reducing the extent of experimental work and as a result reducing the production costs. The methods used for the prediction of mechanical properties of nanocomposites can be classified as (i) theories based on continuum mechanics, (ii) atomistic modeling, and (iii) numerical continuum mechanics [13]. In this study, the finite element method (FEM) based on the numerical continuum mechanics approach has been used for modeling CNT/epoxy nanocomposite.

The experimental study of CNTs reinforced polymers is time consuming and incurs high cost. On the other hand, due to the hypotheses imposed on the micromechanical equations, it is not possible to use them to predict the mechanical properties of nanocomposites containing CNTs nanoparticles. On the other hand, computationally, the use of FEM is superior to the molecular dynamics in terms of solution time, but considering the interface as a continuum medium with homogeneous properties without modeling the interatomic potential of the van der Waals forces is not consistent with the actual behavior of the material.

Odegard et al. [14] introduced a method using a combination of molecular dynamics, a nanostructure, and a continuum mechanics theory to calculate the mechanical properties of nanocomposites. They showed that their method could be used to predict the mechanical properties of polymer nanocomposites with single-wall carbon nanotubes (SWCNTs) in a directional or random orientation. Wan et al. [15] calculated Young's modulus of nanocomposites by considering an isotropic layer at the interface with its Young's modulus a multiple of the polymer matrix Young's modulus. Liu and Chen [16,17] modeled the cylindrical and cubic RVE of nanocomposite using the FEM, assuming prefect bond at the interfacial phase. They considered this volumetric element as reinforcement and used the rule-of-mixtures (ROM); then, the results were compared with the experiments showing that the mechanical properties of the volumetric element created by this method could be reliable. Using a cubic shaped RVE, Fereidoon et al. [18] created a 3D model of CNTs and polymers around them. In their model, CNTs were modeled with the beam elements, showing that the changes in Young's modulus increases linearly with increasing the CNTs volume fraction.

Shokrieh and Rafiee [19] also obtained the Young's modulus of the nanocomposite by using a FEM and modeling van der Waals forces at the interface by using the nonlinear spring elements at the nanoscale. Comparison of the results by rule-of-mixtures showed that this rule did not have the capability to predict the elastic properties of nanocomposites and had an approximation more than the actual values. Shojaie and Golestanian [20] considered a cubic RVE containing a CNT. They simulated the interface as a thin film with elastic properties and compared the results with those of micromechanics. The modeling was done by changing the mechanical properties of the interface region to reach perfect bond. The results are compared with those from rule-of-mixtures showed that the mechanical properties of the nanocomposite were significantly dependent on the surface strength of the interface. In a similar study, Joshi et al. [21] calculated Young's modulus of the nanocomposite by the FEM using a hexagonal RVE. They studied the effect of changing the angle of the CNT on the Young's modulus. They found that the Young's modulus of the RVE was dependent on the angle of the CNTs; by increasing the angle of the CNT relative to the axis of loading, the Young's modulus was decreased. Deng et al. [22] have shown experimentally that the effective elastic modulus of CNT as reinforcement of polymer is far less than the theoretical value. Using a simple tensile test on a nanocomposite film whose particles were polarized by the laser and comparison of the results with the rule-of-mixtures, the value of the effective elastic modulus of the nanoparticles was declared to be between 50% and 70% of the theoretical value.

Peng et al. [23] used a two-scale modeling to create a 2D FEM of nanocomposite with polymer nanoparticles. The interface was considered as an effective interface and according to the elliptical geometry of the nanoparticles, the effect of aspect ratio, radius of the nanoparticle, orientations, volume fraction, and clustering on the Young's modulus of subcell and unit cell were investigated. The results were compared and validated with the results of Mori-Tanaka approach. Eventually, they showed that the elastic modulus increases with increasing the aspect ratio of nanoparticles. Also, they showed the thickness and properties of the interface, the shape of the nanoparticles, their orientation, and degree of clustering have strong influence on the elastic modulus of nanocomposite.

Zuberi and Esat [24] used FEM to model the interface by two perfect bond methods, using truss/spring elements. They found that the nanotube structure affected the properties of the RVE, and the Armchair structure causes a higher increase of properties than Chiral and Zigzag types. Subramanian et al. [25] simulated the traction–separation behavior at the interface in order to achieve force-displacement cohesive behavior by molecular dynamics. By extracting the result from different fracture modes, they developed some analytical equations based on the interface behavior.

Previous studies indicate that the modeling of nanocomposites involves many requirements, and all of them add to the complexity of the modeling and the calculations become complicated and time-consuming. As a result, by selecting the more important and effective parameters, the computational time can be reduced to an acceptable level while still having high accuracy. The interface surface between CNTs and epoxy is one of the major parameters in nanocomposites modeling. Often, molecular dynamics simulation or its results are used to model the interface but it involves a high computational time [26,27]. In some studies, the interface is considered as perfect bond [16–18,28]; however, the interface between the nonfunctionalized CNTs and the surrounding resin in the normal state consists of weak van der Waals bonds. Some scholars have compensated for errors, due to van der Waals bonds at the interface, by considering their properties as a multiple of the properties of CNT and polymer [20,21].

In the present study, the Young's modulus of CNT/epoxy nanocomposite has been evaluated at a multiscale level. The interface behavior is modeled at nanoscale as it is the most significant feature of nanocomposites as compared with the conventional composites. Using the EF technique at this scale can reduce computational costs in simulations at the larger scale. At the microscale, the EF is distributed as reinforcement with properties extracted from the RVE at nanoscale. Due to the fact that the EF outer layer and the matrix of the RVE are of the same material, the bond between them is considered to be perfect. Finally, the simulation results are compared with the experimental results as well as with the results from the Halpin–Tsai model [29,30].

2. Nanocomposite Modeling

From Marino's point of view, important sources of error in simulating the behavior of nanocomposites can be attributed to interfaces, agglomeration, and morphology [13]. The modeling of the interface between resin and nanoparticles is one of the main issues in the present study. Given that the applied CNT in this study is nonfunctionalized, the behavior of the interface is based on van der Waals forces between the nanotubes and polymer molecules; it is defined by the Park–Paulino–Roesler (PPR) potential model and is implemented with user subroutine UEL in the ABAQUS finite element software. On a volumetric scale, the dispersion pattern for nanofibers is taken randomly.

2.1. Interface

In conventional composites, the forces between the matrix and the fibers account for a small proportion of the interatomic forces in the structure of the composite. But in nanocomposites, where the surface-to-volume ratio is much higher, these forces are more important. Therefore, a model that can simulate the surface and the interatomic forces of the surface more precisely is appropriate [31].

In the current research, cohesive elements are used to model the interface behavior. The behavior of these elements is defined based on a nonlinear traction–separation relationship. From an atomistic perspective, failure occurs when the separation between atoms occurs.

Figure 1 shows a schematic of the behavior between the two atoms and the energy required for their separation [32]. Accordingly, the cohesive force between the atoms acts in a nonlinear manner in separation and the assumption of the linear behavior for this force is not correct.

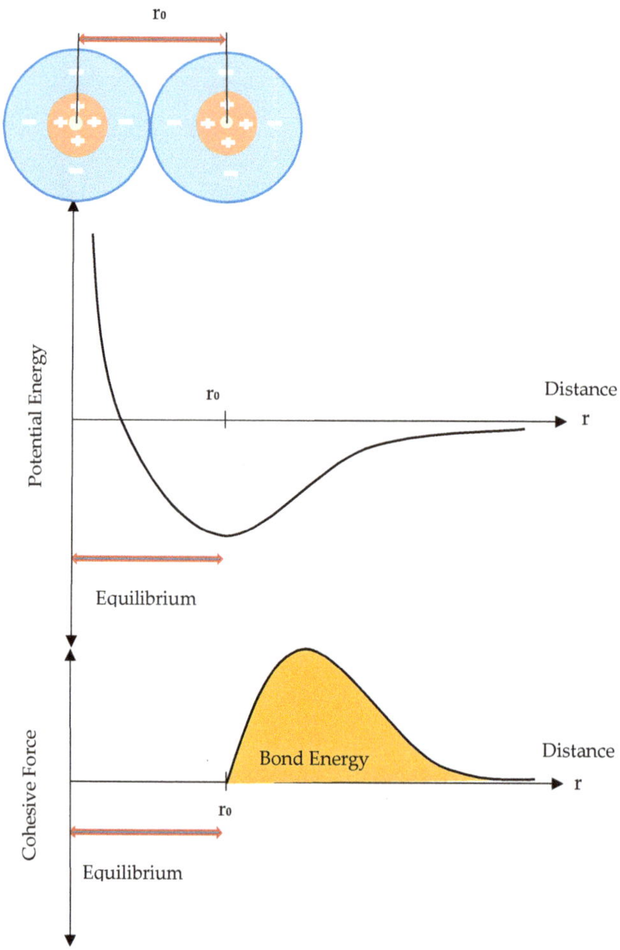

Figure 1. Variation of the potential energy and the cohesive force during separation of two atoms from each other.

Carbon nanotubes are cylindrical molecules consisting of a circular array of sp^2 hybridized carbon atoms, which makes it impossible to form a covalent bond between the nanotube carbon atoms, and the bond with the surrounding polymer molecules can be achieved through weak van der Waals forces [25,33]. Given the fact that the energy contribution of van der Waals interactions in the intermediate phase is three times higher than the electrostatic bond energy, the electrostatic bonds are ignored compared to the van der Waals bonds [34]. The bond between the carbon atoms in the outermost layer of the untreated CNT and the matrix molecules on the interface is formed by van der

Waals bonds, which can be defined by the Lennard–Jones (L–J) Potential function. The L–J Potential is the most commonly used expression for the attraction/repulsion properties of the interaction between atoms and neutral molecules, expressed by Equation (1) [31].

$$V_{LJ}(r) = 4\varepsilon\left[\left(\frac{\sigma}{r}\right)^{12} - \left(\frac{\sigma}{r}\right)^{6}\right] \tag{1}$$

r is the distance between the atoms, and ε and σ are the van der Waals parameters, which are 0.4492 kJ/mol and 0.34 nm for carbon atoms, respectively. The variation of the Lennard–Jones force in terms of the distance between the carbon atoms is shown in Equation (2) [35]:

$$F_{LJ}(r) = -24\frac{\varepsilon}{\sigma}\left[2\left(\frac{\sigma}{r}\right)^{13} - \left(\frac{\sigma}{r}\right)^{7}\right] \tag{2}$$

The van der Waals force versus interatomic distance as well as force versus displacement are depicted in Figure 2 [36,37]. Jiang et al. [38], Tan et al. [39], and Zakeri et al. [36] are among the researchers used the L–J Potential to calculate the cohesive force at the interface between CNTs and polymer matrices.

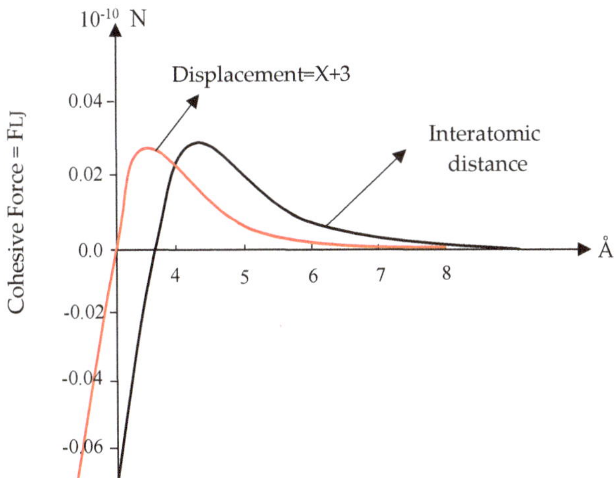

Figure 2. The van der Waals force versus interatomic distance and displacement curve.

The equilibrium distance in Equation (1) for carbon atoms is $r_0 = 3.8$ Å, and for the calculation of forces in terms of displacement, we can rewrite Equation (2) in the form of Equation (3). In this equation, x is the amount of atomic displacement [30].

$$F_{LJ}(r) = -24\frac{\varepsilon}{\sigma}\left[2\left(\frac{\sigma}{x+3.8}\right)^{13} - \left(\frac{\sigma}{x+3.8}\right)^{7}\right] \tag{3}$$

2.2. Potential Model

Until now, there have been different equations for force-displacement between the adhesive surfaces, but none of them are capable of defining the nonlinear behavior on the adhesive surface completely. As a result, there is a difference between the actual behavior of the interface with these models. Park, Paulino, and Roesler (PPR) derived the relation between the potential sticky surfaces as presented in the Equation (4) [40].

$$\Psi(\Delta_n, \Delta_t) = \min(\phi_n, \phi_t) + \left[\Gamma_n(1 - \tfrac{\Delta_n}{\delta_n})^{\alpha}(\tfrac{m}{\alpha} + \tfrac{\Delta_n}{\delta_n})^{m} + \langle\phi_n - \phi_t\rangle\right] \times \left[\Gamma_t(1 - \tfrac{|\Delta_t|}{\delta_t})^{\beta}(\tfrac{n}{\beta} + \tfrac{|\Delta_t|}{\delta_t})^{n} + \langle\phi_t - \phi_n\rangle\right] \tag{4}$$

In Equation (4) δ_n, δ_t, Γ_n, and Γ_t are:

$$\delta_t = \frac{\phi_t}{\tau_{\max}}\beta\lambda_t(1-\lambda_t)^{\beta-1}(\frac{\beta}{n}+1)(\frac{\beta}{n}\lambda_t+1)^{n-1} \tag{5}$$

$$\delta_n = \frac{\phi_n}{\sigma_{\max}}\alpha\lambda_n(1-\lambda_n)^{\alpha-1}(\frac{\alpha}{m}+1)(\frac{\alpha}{m}\lambda_n+1)^{m-1} \tag{6}$$

$$\Gamma_n = (-\phi_n)^{\frac{\langle\phi_n-\phi_t\rangle}{(\phi_n-\phi_t)}}(\frac{\alpha}{m})^{m}, \Gamma_t = (-\phi_t)^{\frac{\langle\phi_t-\phi_n\rangle}{(\phi_t-\phi_n)}}(\frac{\beta}{n})^{m} \tag{7}$$

There are eight input parameters in the PPR potential model: modes I and II fracture energies (ϕ_n, ϕ_t), normal and tangential cohesive strengths ($\sigma_{\max}$, $\tau_{\max}$), normal and tangential shape parameters (β, α), and the initial slope indicators (λ_t, λ_n), according to Figure 3. The characteristic length scale parameters (δ_n, δ_t) is given by the Equations (5) and (6), as well as their relation with other parameters [40,41].

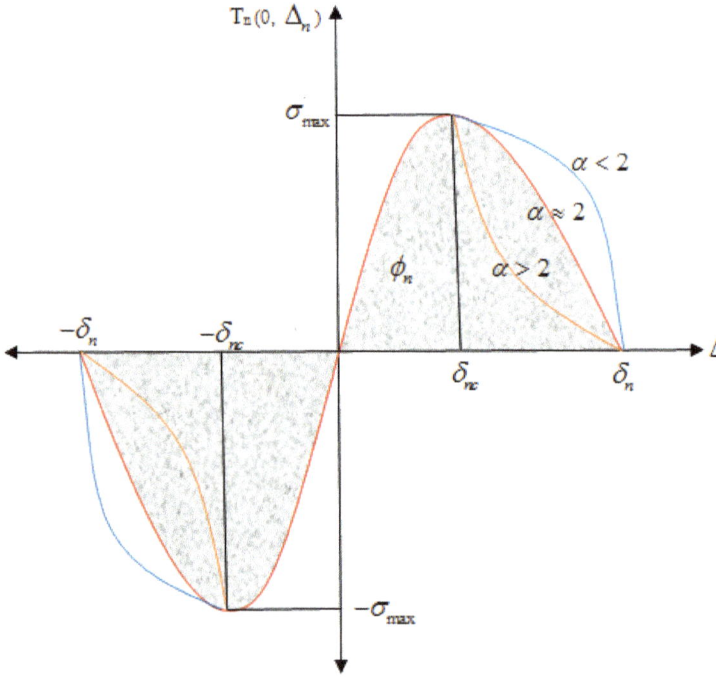

(a)

Figure 3. *Cont.*

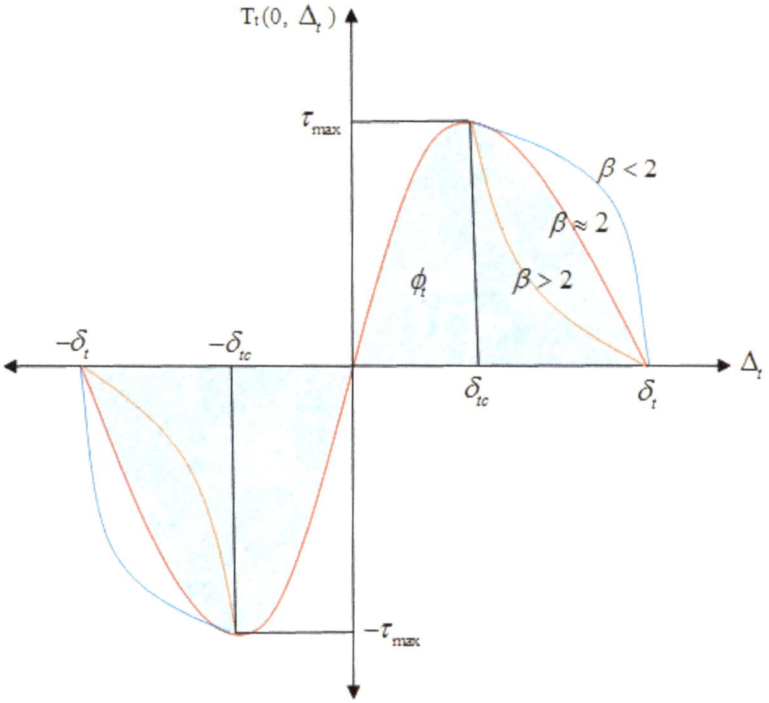

(**b**)

Figure 3. Traction–separation diagram in the (**a**) normal and (**b**) tangential mode [40,42].

Appendix A describes how to calculate these parameters in the tangential mode [35]. If the energy values of the normal and tangential fracture are equal, then Equation (7) becomes the same as Equation (8). m and n are calculated by means of the Equation (9), based on the shape parameters in the PPR model (α, β) and the initial slope indicators, λ_n and λ_t which are defined by the Equation (10).

$$\Gamma_n = -\phi\left(\frac{\alpha}{m}\right)^m, \Gamma_t = \left(\frac{\beta}{n}\right)^n \quad (8)$$

$$m = \frac{\alpha(\alpha-1)\lambda_n^2}{(1-\alpha\lambda_n^2)}, n = \frac{\beta(\beta-1)\lambda_t^2}{(1-\beta\lambda_t^2)} \quad (9)$$

$$\lambda_n = \frac{\delta_{nc}}{\delta_n}, \lambda_t = \frac{\delta_{tc}}{\delta_t} \quad (10)$$

Figure 3 schematically shows the traction–separation diagram in the normal and tangential mode based on the PPR potential model [40]. In Figure 4, the traction–separation diagram is shown in the mixed mode. Since there is a good agreement between the Lennard–Jones variation diagram in Figure 2 and the PPR model in Figure 4, this potential can be used as the basis for defining the adhesive properties in the form of nonlinear elasticity at the interface, whose main parameters values are based on the L–J potential.

The strain energy of interface on a continuous surface can be calculated using Cauchy–Born rule as shown in Equation (11).

$$\phi = \frac{\int V_{LJ}(r)dr}{S_0}, \left(S_0 = \frac{3\sqrt{3}}{4} r_{ij}^2\right) \quad (11)$$

where r_{ij} is the distance between carbon atoms [43,44]. Using this idea makes it possible to simulate the interface by considering its actual behavior. The potential model of the Equation (4) is implemented for use in the finite element software and coded in the Fortran [45]. The eight input parameters for PPR potential model are calculated by Equations (4)–(11), Figure 2, and Appendix A listed in Table 1.

Table 1. The values of input parameters to the Park, Paulino, and Roesler (PPR) model.

Value	Input Parameter
$\phi_n = 0.2916081e^{-7}$ $\phi_t = 0.1241959e^{-7}$	Fracture energy (N/nm)
$\sigma_{\max} = 1.38861e^{-8}$ $\tau_{\max} = 0.591409e^{-8}$	Cohesive strength (N/nm^2)
$\alpha = 5$, $\beta = 5$	Shape parameters
$\lambda_n = 0.0851$, $\lambda_t = 0.23516$	Initial slope indicators

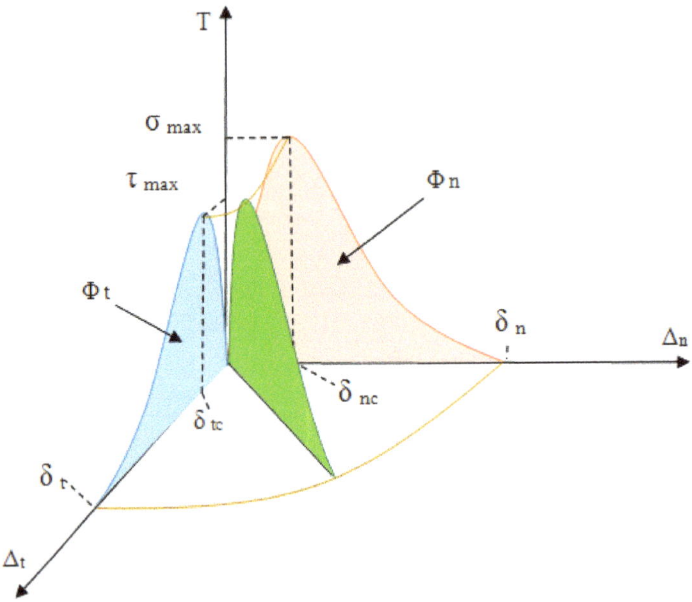

Figure 4. Normal and tangential mixed-mode response of the cohesive element in a three-dimensional state.

2.3. Equivalent Fiber

The use of EF technique is an efficient method for reducing the volume of computation. By using this method, the behavior of nanocomposites is similar to that of conventional composites [16,20,21,27]. It should be noted that the structure of the epoxy matrix at the vicinity of the CNT surface differs from the bulk epoxy due to the formation of an ultrathin epoxy layer at the CNT/epoxy interface. This ultrathin layer consists of highly packed crystalline epoxy monomers, which has different elastic properties in comparison with the amorphous bulk polymer [46,47]. It is shown that the radial distribution function (RDF) of epoxy atoms is zero at the radial distance of 0.56 nm and reaches its maximum value of 160 atoms/nm^3 at the radial distance of 0.77 nm [48]. Then, it starts to fluctuate around an average value of 110 atoms/nm^3. This result indicates that the value of the van der Waals equilibrium distance is ~2.75 Å and the thickness of CNT/epoxy matrix interphase layer is ~3.0 Å.

The schematic of the equivalent fiber model is shown in Figure 5. The mechanical properties of the nanotubes are based on the material which is used in experimental tests, and is shown in Table 2 [49,50]. The multiwalled carbon nanotubes (MWCNTs) are made by the USNANO company, with an inside diameter of 5 to 10 nm, an outer diameter of 10 to 20 nm, and a length of 10 to 30 μm. In the modeling, the CNTs' average internal diameter of 7.5 nm and the external diameter of 15 nm were assumed. KER828 epoxy with the Young's modulus of 2.6 GPa, Poisson's ratio of 0.3, and density of 1.16 g/cm^3 was chosen. The resin Young's modulus was obtained through tensile test according to the ASTM-D638 standard. Due to the high aspect ratio of MWCNT, the modeling of CNTs in the RVE was subjected to limitations. In order to create the finite element model, researchers have considered the aspect ratio between 6 and 70 [49].

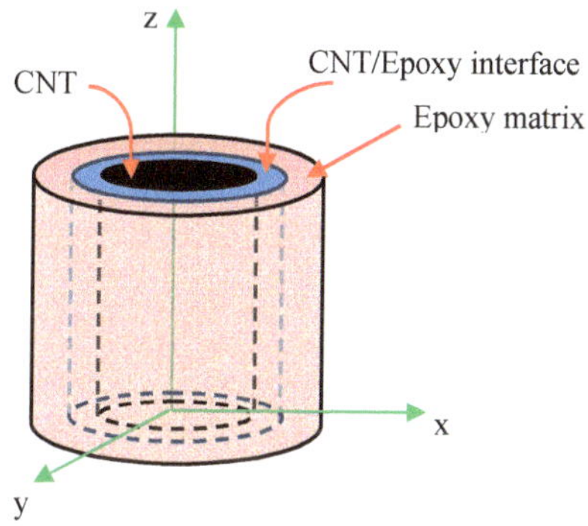

Figure 5. A schematic of the equivalent fiber model.

Table 2. Mechanical properties of multiwalled carbon nanotube [49,50].

Transversely Isotropic	
Young's modulus (E_z)	1 TPa
Young's modulus (E_x = E_y)	30 GPa
Shear modulus (G_{xy} = G_{xz} = G_{yz})	30 GPa
Poisson's ratio (ν_{xy} = ν_{xz} = ν_{yz})	0.0687
Density (ρ)	2.1 g/cm^3

3. Simulation of Nanocomposite Behavior

The high computational costs for atomic and molecular dynamics modeling of the behavior of nanocomposites at a nanoscale, as well as the limitations of these methods in higher scales, and the difference between the results of micromechanical theories and the experimental ones have all led researchers to consider alternative methods. In fact, the difference between atomic behaviors and continuum mechanics hypotheses is an obstacle to the use of these methods. The use of multiscale methods and the transfer of properties from smaller scale to a higher one with the aim of reducing the computational time have been considered in many studies employing the FEM and atomistic methods [27,35,37,48,51–53]. In this study, the multiscale approach is used for modeling atomic behavior at the interface. Finite element simulation is used at nanoscale and microscale for the prediction of the nanocomposite Young's modulus. At the nanoscale, the EF was simulated and its mechanical properties were extracted and transferred to microscale model.

3.1. Equivalent Fiber Simulation

In this section, the finite element is simulated according to Section 2.3. Since the ABAQUS software library does not include cohesive elements with the nonlinear elastic behavior, a user element subroutine (UEL) was written to define the interface behavior based on the PPR potential model. The input parameters of the model were extracted and applied, using the L–J atomic potential by using Equations (1)–(4), as shown in Figure 2. The finite element simulations of the uniaxial tensile, transverse loading, and axial torsion tests have been performed. From finite element analysis (FEA) results and Equations (12)–(17), the mechanical properties of the EF, including transverse and longitudinal Young's modulus, the shear modulus, and Poisson's ratio were extracted. Figure 6 illustrates the steps of the modeling at nanoscale.

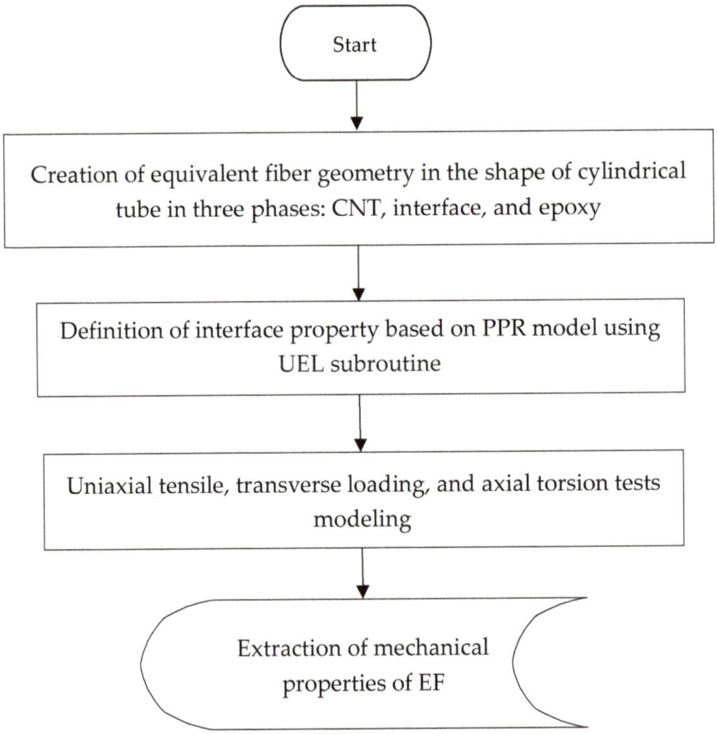

Figure 6. Modeling steps at the nanoscale.

The EF was simulated in the uniaxial tensile loading as shown in Figure 7, and the reaction force at the support is computed and longitudinal Young's modulus, E_z, has been extracted from Equation (12) and Poisson's ratio, ν_{zx}, from Equation (13). Moreover, in the transverse loading, the EF has been constrained at both ends along its axial direction to make a plane strain condition; then, Equation (14) was used for calculation of the transverse Young's modulus. Also, by applying torsional loading, variation of the twist angle in the EF was computed and the shear modulus was found using Equation (15). By solving Equations (14) and (15), the Young's modulus, E_x, and Poisson's ratio, υ_{xy}, can be obtained as shown in Equations (16) and (17). In this work, the study is focused on the effect of CNTs on the Young's modulus of the epoxy in the elastic limit. For investigating the strength of the CNT reinforced epoxy nanocomposite at a microscale, the effect of interfacial failure and the local damage should be considered. The distributions of von Mises stress and mid principal stress of EF under uniaxial tensile loading are shown in Figure 8.

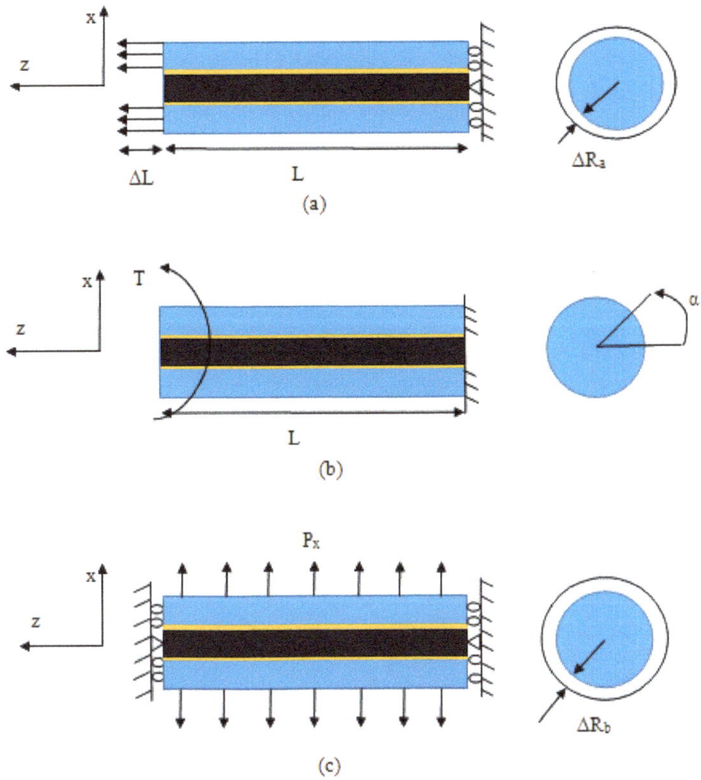

Figure 7. Performing (**a**) axial tensile; (**b**) torsional; and (**c**) transverse tests in simulation.

Calculation of the longitudinal Young's modulus:

$$E_z = \frac{P_z L}{A \Delta L} = \frac{\sigma}{\varepsilon} \tag{12}$$

Calculation of the Poisson's ratio between the longitudinal axis and the transverse plane:

$$v_{zx} = \frac{-\left(\frac{\Delta R_a}{R}\right)}{\left(\frac{\Delta L}{L}\right)} \tag{13}$$

Elastic equation for transverse direction:

$$-\left(\frac{v_{xy}}{E_x} + \frac{{v_{zx}}^2}{E_z}\right) + \left(\frac{1}{E_x} - \frac{{v_{zx}}^2}{E_z}\right) = \frac{\Delta R_b}{P_x R} \tag{14}$$

Calculation of the shear modulus:

$$\frac{TL}{\alpha J} = G_{xy} = \frac{E_x}{2(1 + v_{xy})} \tag{15}$$

Young's modulus, E_x, and Poisson's ratio, v_{xy}, are calculated from Equations (16) and (17):

$$E_x = E_y = \frac{2}{\frac{1}{2G_{xy}} + \frac{2v_{zx}^2}{E_z} + \frac{\Delta R_b}{P_x R}} \quad (16)$$

$$\nu_{xy} = \frac{E_x}{2G_{xy}} - 1 = \frac{1}{G_{xy}\left(\frac{1}{2G_{xy}} + \frac{2v_{zx}^2}{E_z} + \frac{\Delta R_b}{P_x R}\right)} - 1 \quad (17)$$

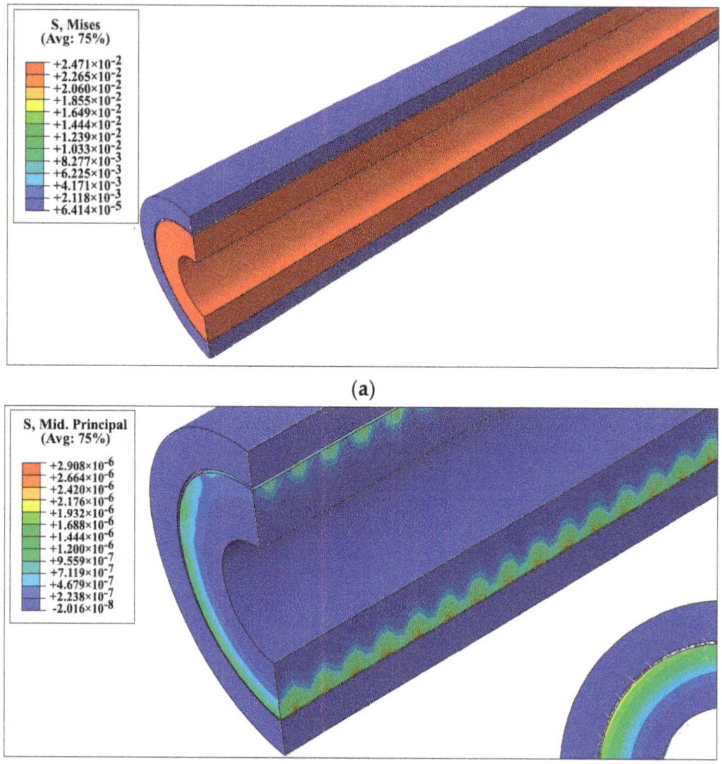

(b)

Figure 8. Results of finite element simulation of equivalent fiber (EF) under uniaxial loading (**a**) contour plots of von Mises stress (**b**) mid-principal stress distribution.

The actual length of the CNTs is based on the production process, the arrangement of atoms, and the number of walls, ranges from 100 nm to 15 μm. The dispersion of the EFs in the RVE with the actual length of the CNTs is awkward.

Due to the aspect ratio of CNTs (~700), the equivalent fiber boundary conditions are defined to model an infinite length for the carbon nanotube. Figure 9 illustrates variation in the longitudinal and transverse Young's modulus with respect to the change in EF length from 50 nm to 600 nm. This diagram shows that the longitudinal Young's modulus does not change significantly for EF lengths around 300 nm, and for fibers length over 300 nm, the change in transverse Young's modulus of EF changes is negligible; hence, in the modeling of EF its lengths was considered equal to 300 nm.

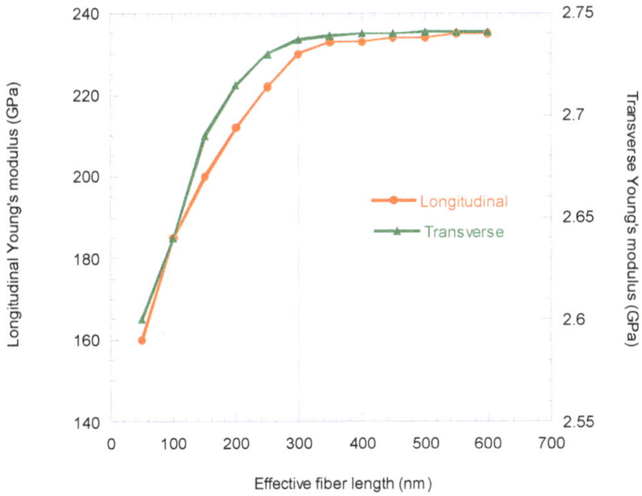

Figure 9. Variation in the longitudinal and transverse Young's modulus with respect to the change in EF length.

In Figure 10, the longitudinal stress–strain curve obtained from the FEM for EF is compared with the results of the rule-of-mixtures. The results show that up to the strain of 0.045 the behavior is linear. The results obtained for the mechanical properties of the EF also shows that the EF behavior is transversely isotropic, and this behavior was due to the behavior of the interface and the transversely isotropic nature of CNTs.

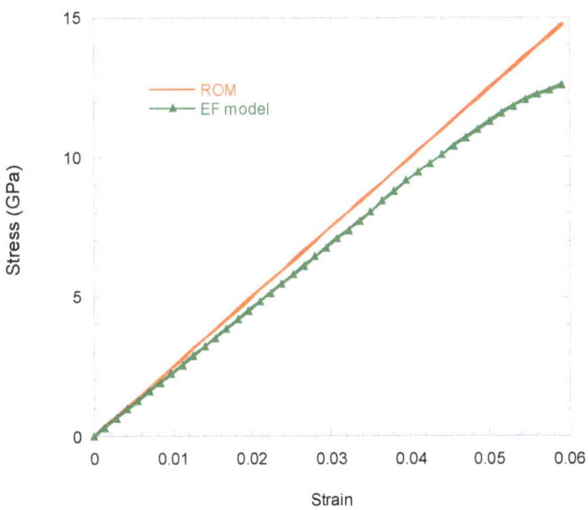

Figure 10. Comparison of longitudinal stress–strain behavior of the finite element analysis EF model and the rule-of-mixtures (ROM).

3.2. RVE Simulation

This study has been done for volume fractions of less than 1%, and dispersion is done by ultrasonic energy. Reducing the volume fraction of CNTs has a direct effect on the agglomeration [10].

The agglomeration of CNTs is ignored in this section and the effect of CNTs clustering by the FEM is investigated in the next section. In the simulation of the RVE, it is assumed that the bond between the EF and the matrix is a perfect bond and the position and angle of the reinforcement are chosen randomly. This model is implemented by the Python script in the ABAQUS, which allows the script to select and modify the properties and dimensions of the matrix and the reinforcement, and the values and properties of the EF are based on nanoscale outputs. Figure 11 shows the RVE with dimensions of $(1500 \times 1500 \times 1500)$ nm^3. The Young's modulus of the RVE was calculated using Equation (18). In Figure 12 the algorithm of the RVE simulation is shown. The results of this model, as the full dispersion model, are presented in Section 5.

$$E = \frac{PL}{A\Delta L} = \frac{P}{A\varepsilon} \tag{18}$$

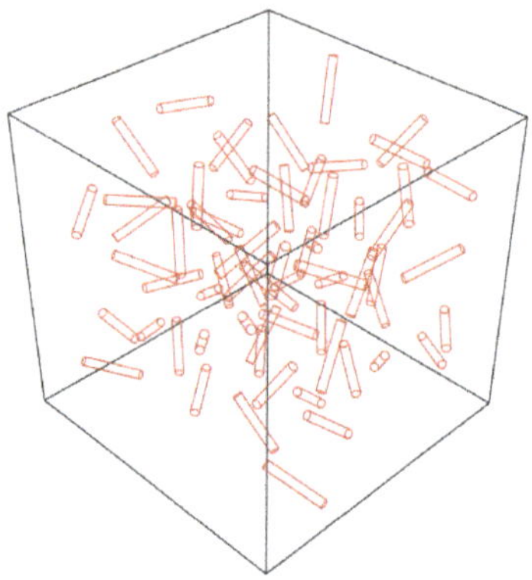

Figure 11. 3D view of the representative volume element (RVE) with the volume of $1500 \times 1500 \times 1500$ nm^3 and the random dispersion of the equivalent fibers.

3.3. *Simulation of CNTs Agglomeration*

Incorporation of CNTs can improve significantly the mechanical properties of epoxy at <3 wt% loading if they are properly dispersed and good interfacial bonding between the CNTs and polymer matrix is achieved. However, due to strong intermolecular van der Waals interactions, CNTs tend to form ropes or bundles. These bundles can aggregate further, forming entangled networks or agglomerates; hence deteriorating the mechanical properties of nanocomposite as contribution from individual CNTs cannot be optimally exploited. One of the methods for improving dispersion and interfacial adhesion of CNTs is amino-functionalization. Typically, carboxylic groups (–COOH) are produced by treating CNTs with high concentration acids, followed by further functionalization using amine molecules. Figure 13 compares dispersion of pristine CNT and aminofuntionalized-CNT in an epoxy resin [54–56]. Therefore, it is important to fully understand the effect of agglomeration of CNTs on mechanical behavior of nanocomposite.

In this section, the effect of agglomeration on the Young's modulus of nanocomposite has been investigated by FEM. The RVE is divided into 64 sub cubes and half of the CNTs are clustered in 1, 2, and 4 colonies, respectively, and the rest are dispersed randomly in RVE. Figure 14 shows a 3D view of the RVE with four colonies in different volume fractions. For a volume fraction of 1%, which is most effective against agglomeration, the effect of the ratio of nanotubes aggregated to the whole nanotube in RVE is investigated in four states of 25%, 50%, 75%, and 100%. The results of these models, as the agglomeration model, are presented in Section 5.

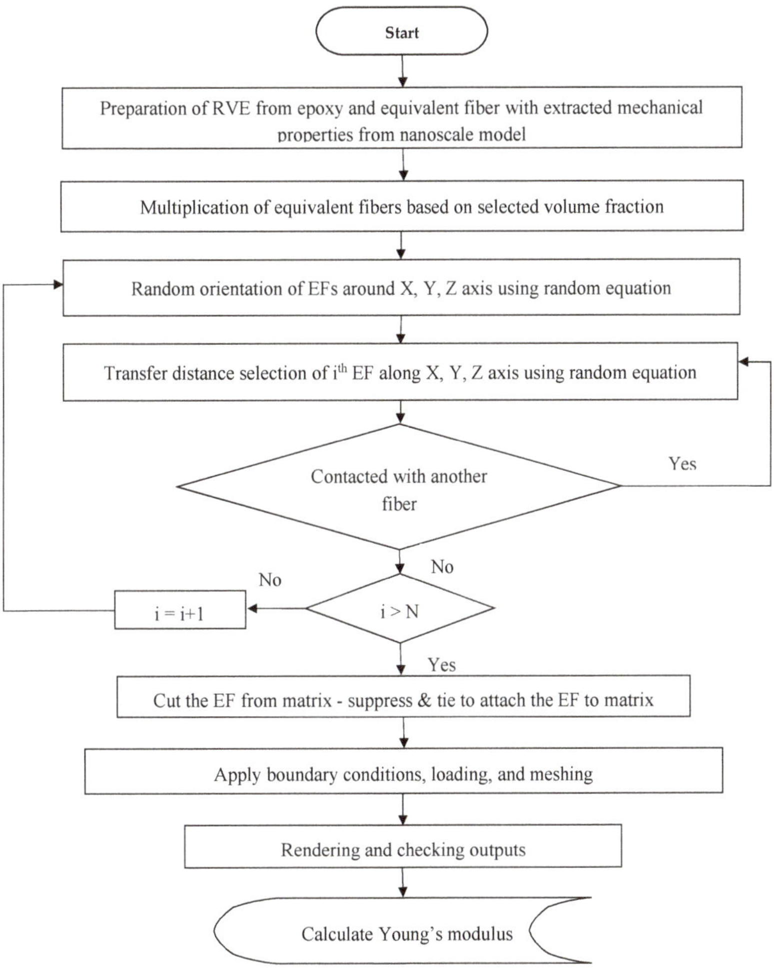

Figure 12. Flowchart of the RVE modeling at the microscale.

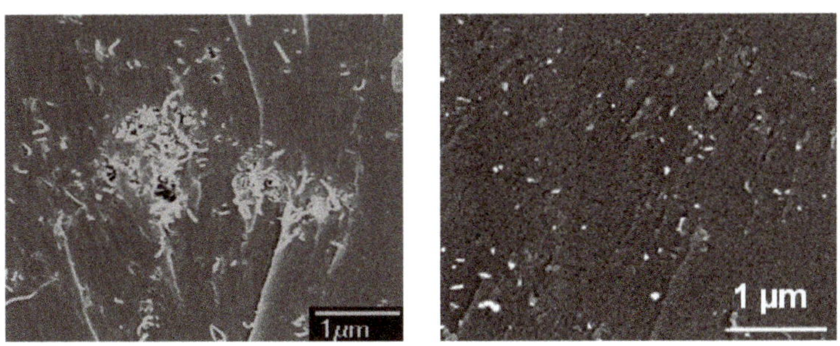

Figure 13. SEM images of fracture surface of epoxy containing 0.5 wt.% pristine carbon nanotubes (CNTs) (**left**) and amino-CNTs (**right**). Reproduced with permission from [54]. Elsevier, 2010.

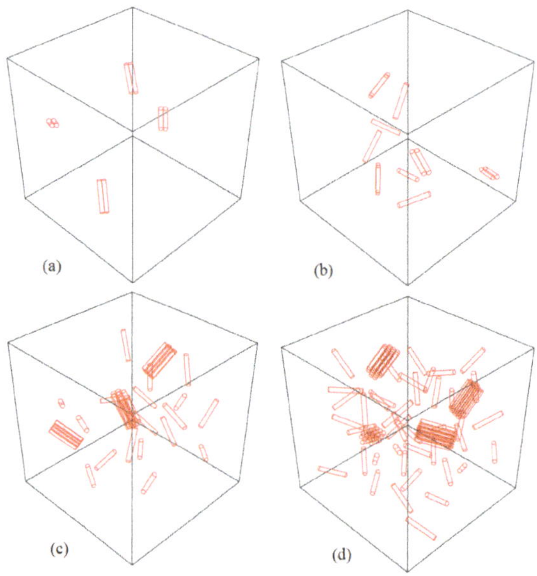

Figure 14. 3D view of the RVE with clustering of CNTs in four colonies: (**a**) 0.18 wt%; (**b**) 0.36 wt%; (**c**) 0.9 wt%; (**d**) 1.8 wt% (half of the EFs are clustered at four random point).

4. Experimental Studies

Nanocomposite specimens were made and tested in accordance with the ASTM-D638 standard Type V. The production of the samples was done using multistage production instructions aimed at the highest degree of separation and dispersion. Silicon was used for molding, which is a thermosetting polymer group with good flexibility, allowing for the removal of specimens without any damage. Acetone was used as a solvent for better dispersion of MWCNTs in the epoxy.

Table 3 shows the amount of each of the constituents' materials for different volume fractions of CNTs.

Table 3. Ratios of materials used for the production of nanocomposites with different volume fractions of multiwalled carbon nanotubes (MWCNTs).

Acetone (mL)	MWCNT (mg)	Hardener (g)	Epoxy (g)	CNT Vf.%	CNT wt.%
30	30	1.5	15	0.1	0.18
60	60	1.5	15	0.2	0.36
150	150	1.5	15	0.5	0.9
300	300	1.5	15	1	1.8

Volume fraction % = Vf.%; Weight fraction% = wt.%.

Figure 15 illustrates the process of nanocomposite specimens' production. To make the specimen, the CNT was first mixed with acetone and placed in an ultrasonic bath for an hour. After that, the epoxy was added and placed in an ultrasonic bath for another hour. This process dispersed CNTs in the epoxy. In the next step, for the removal of acetone and air bubbles, the nanocomposite was placed in a vacuum oven for 10 h (to be sure, the mass of the mixture was measured after removal from the oven). Then, by adding the hardener, the mixture was slowly mixed for 5 min and placed in a vacuum oven for 10 min. At the end, the nanocomposite was poured into the dogbone mold and after 24 h, the specimens were ready for tensile testing. Tensile test was conducted according to the ASTM-D638V standard with the universal testing machine SANTAM-STM50. The values of the Young's modulus obtained from this experiment were compared with the results of the simulation and the Halpin–Tsai equation in Figure 16.

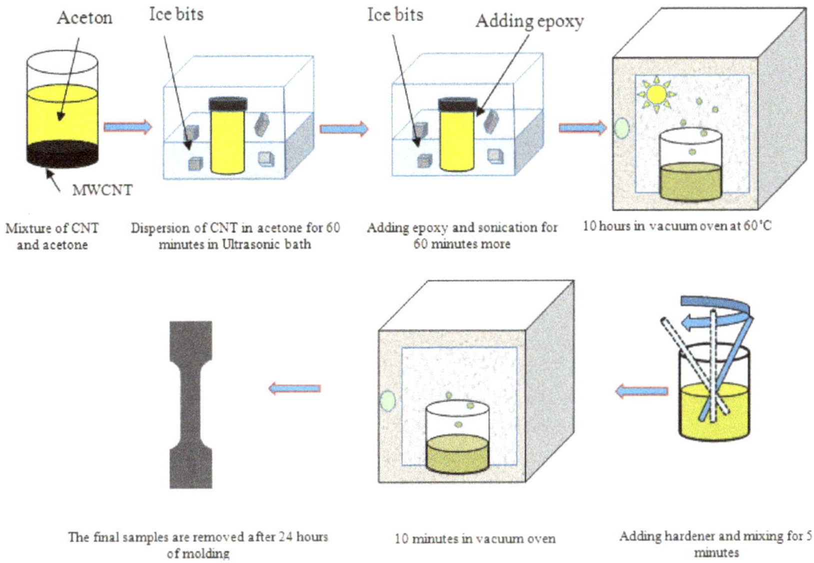

Figure 15. The production stages of CNT/epoxy nanocomposite.

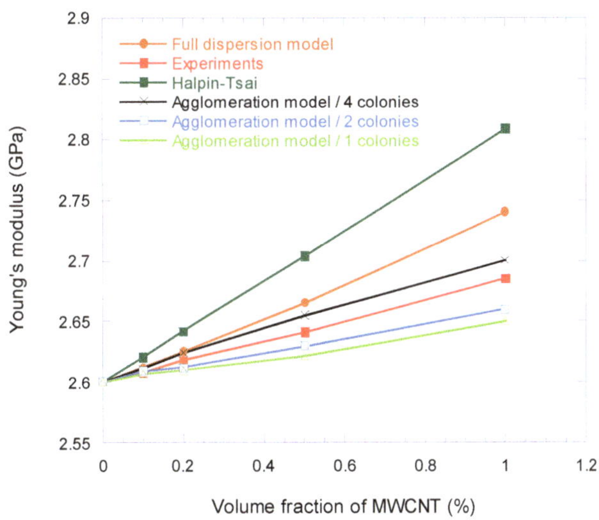

Figure 16. Variation of Young's modulus of CNT/epoxy nanocomposite with respect to MWCNT volume fraction.

5. Result and Discussion

According to the results shown in Figure 10, for EF length longer than 300 nm, the longitudinal and transverse EF Young's modulus does not depend on its length. A longitudinal Young's modulus of 230 GPa and transverse Young's modulus of 2.74 GPa were obtained. Although the linear behavior was considered to be for CNTs and epoxy, a nonlinear behavior is exhibited at strains greater than 0.045 of the EF, which is similar to the nature of the van der Waals bonds at the interface.

The FEA simulation results are compared with Halpin–Tsai model defined in Equations (A5)–(A9) in Appendix B, along with the experimental results. The results of nanocomposite Young's modulus form experiment, Halpin–Tsai and full dispersion FEA simulation together with agglomerated CNTs FEA models for 1, 2, and 4 colonies are compared in Figure 16. The results show that the values obtained from simulations are between the results from Halpin–Tsai equation and experiments and the agglomerated CNTs FEA model with four colonies is the closest to the experiment. This shows that the dispersion of CNTs in the epoxy was not perfect and there were some agglomerations.

The full dispersion model showed that for MWCNTs loading at 1 Vf.%, the Young's modulus of epoxy nanocomposite was increased by 5%. This number was 1.7% more than the experimental value, while the result from the Halpin–Tsai equation overestimated 4.6% the experimental value at the same volume fraction. If the results are compared with the rule-of-mixtures, the difference will be more than 20%. The proximity of the results of the FEA modeling with the experimental ones showed the effect of interface modeling on the elastic behavior and the Young's modulus of the nanocomposite. In Figure 17, the normalized differences of Young's modulus, which are derived from the FEA simulation and Halpin–Tsai equation, are compared with the experimental results. As can be seen in this figure, by increasing the amount of the CNT volume fraction, the difference between the experimental results and the simulation increases and the highest difference at the volume fraction of 1% is 2.05%. However, this deviation is approximately half less than the difference of Halpin–Tsai equation and the experiment. The augmentative growth of normalized difference between the full dispersion model and experiment by increasing the volume fraction of CNT is rooted in agglomeration at higher loading of the CNT nanoparticles. Figure 18. shows the effect of clustered CNTs percentage (in four colonies) in a 1% volume fraction on Young's modulus. As can be seen, the increase in agglomerated CNTs volume fraction decreases the modulus of the nanocomposite.

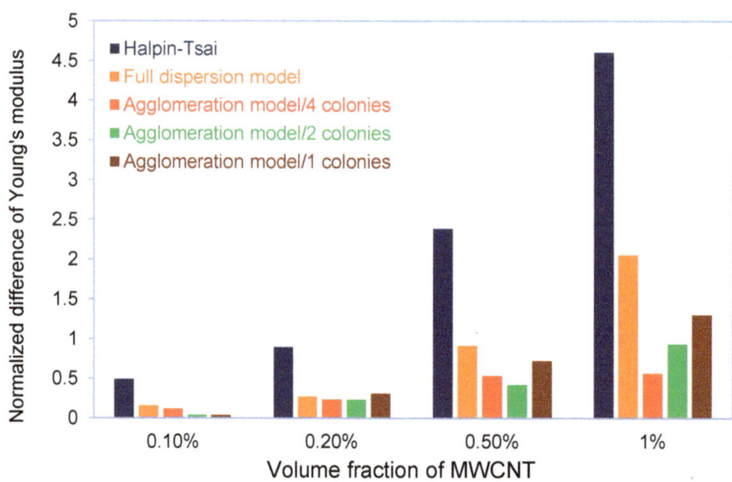

Figure 17. Comparison of the normalized difference of Young's modulus from FEA modeling and Halpin–Tsai equation relative to the experimental results.

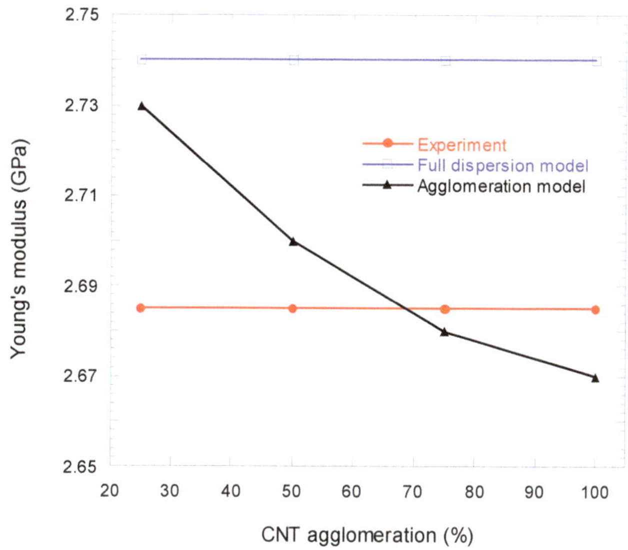

Figure 18. Effect of CNTs agglomeration on Young's modulus of CNT/epoxy nanocomposite for four colonies model and 1 Vf.% CNT loading.

Using the PPR potential-based cohesive model made the results of simulation approach the experimental results closely, which was very accurate in comparison with micromechanical theories. Hence, the nonlinear elastic model could be used to define the adhesive behavior between atomic surfaces, and simulate the proper pattern of interface behavior in the nanocomposites.

6. Conclusions

In this study a CNT/epoxy nanocomposite has been simulated to investigate the elastic behavior of the nanocomposite in tension. At the nanoscale, the interface has been modeled by nonlinear cohesive element. The atomistic behavior of Lennard–Jones is linked with the Park–Paulino–Roesler (PPR) model using Cauchy–Born rule. The simulations have been done based on two-scale method and equivalent fiber technique is used to decrease the amount of computation, by transferring the mechanical properties from nano- to microscale. In the model, the EF is dispersed in the RVE randomly, and Young's modulus of nanocomposites has been extracted from simulations. Also the effect of clustering of CNTs in the matrix is investigated. Both experimental tensile test and Halpin–Tsai equation are used to compare and validate the results of simulation.

The results from this study illustrate that:

1. The PPR model is reliable for the modeling of interface in nanocomposites.
2. The results of modeling at the nanoscale showed that the EF length and the matrix thickness has a diminishing effect on elastic behavior of EF for EF length more than 300 nm. Due to the limitations in simulation of RVE, the minimum length of EF is considered 300 nm.
3. The results from this study have the proximity to experimental results and they predict more accurately than the calculated ones by Halpin–Tsai equation, especially at higher CNT loading.
4. Although some methods were used to disperse CNTs in the production of specimens, the increase in volume fraction has increased the differences between the numerical and experimental results. However, the maximum difference in Young's modulus from the full dispersion model relative to experiment at the volume fraction of 1% is two-third lower than that between Halpin–Tsai equation and experiments.

5. The clustering of equivalent fibers increases the heterogeneity of the RVE, and by increasing the interaction between the equivalent fibers, their contribution to interacting with the environment is low, thereby reducing the properties of the nanocomposite.
6. The agglomeration of CNT in the nanocomposite degrades its effective elastic properties and the effective elastic properties decrease with the increase of the CNT agglomerate size.

Author Contributions: M.J., M.M., and H.H.: Methodology, validation, formal analysis; M.J.: software, and original draft preparation; M.J. and M.K. conceived and designed the experiments; M.J., H.H., and M.M. analyzed the data; M.M. and H.H.: conceptualization and reviewed and edited the manuscript; M.M. supervision and project administration.

Funding: This research received no external funding.

Acknowledgments: The authors wish to thank Polymer and Nanocomposite Labs of IUT for preparing of samples and Strength of Material Lab of IUT for mechanical testing. The authors are also thankful to the National High-Performance Computing Center of Isfahan University of Technology for providing a supercomputer for finite element analysis.

Conflicts of Interest: The authors declare no conflicts of interest.

Appendix A

To use three-dimensional PPR model in addition to normal interface behavior, shear behavior should be determined at this surface. Figure A1 shows how to calculate the tangential component of the van der Waals forces. By using the Equations (A1)–(A3) and substituting them in Equation (11), the tangential force at the interface is obtained using the Lennard–Jones (L–J) Potential, as shown in Equation (A4).

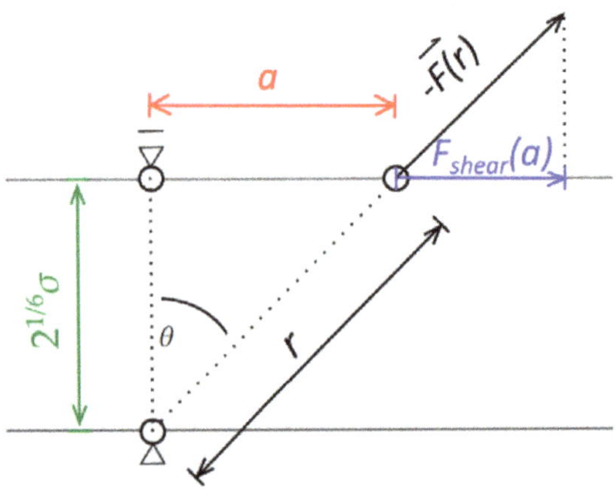

Figure A1. Tangential force component in atomic separation.

$$F_{shear} = F_r \cos\theta \tag{A1}$$

$$\theta = \cos^{-1}\left(\frac{\sqrt[6]{\sigma}}{r}\right) \tag{A2}$$

$$r^2 = a^2 + (\sqrt[6]{2}\sigma) \tag{A3}$$

$$F_{shear}(a) = -4\frac{\varepsilon a}{a^2 + 2^{1/3}\sigma^2}\left[12\frac{\sigma^{12}}{(a^2 + 2^{1/3}\sigma^2)^6} - 6\frac{\sigma^6}{(a^2 + 2^{1/3}\sigma^2)^3}\right] \tag{A4}$$

Appendix B

The Equation (A5) shows how to calculate the Young's modulus of the homogeneous fiber nanocomposite based on the Halpin–Tsai model. The parameters E_{11} and E_{22} were calculated by the Equations (A6) and (A7). In these equations, m is related to matrix and the index f is related to the fiber. E_L and E_T are the longitudinal and transverse reinforcement Young's modulus.

$$E_c = \frac{3}{8}E_{11} + \frac{5}{8}E_{22} \tag{A5}$$

$$E_{11} = E_m\left(\frac{1 + \frac{L}{R_{CNT}}\eta_1 v_f}{1 - \eta_1 v_f}\right) \tag{A6}$$

$$E_{22} = E_m\left(\frac{1 + 2\eta_2 v_f}{1 - \eta_2 v_f}\right) \tag{A7}$$

$$\eta_1 = \frac{\frac{E_{longitudinal}}{E_m} - 1}{\frac{E_{longitudinal}}{E_m} + \frac{L}{R_{CNT}}} \tag{A8}$$

$$\eta_2 = \frac{\frac{E_{Transverse}}{E_m} - 1}{\frac{E_{Transverse}}{E_m} + \frac{L}{R_{CNT}}} \tag{A9}$$

References

1. Plueddemann, E.P. *Interfaces in Polymer Matrix Composites: Composite Materials*; Elsevier: New York, NY, USA, 2016; Volume 6.
2. Kausar, A.; Rafique, I.; Muhammad, B. Review of applications of polymer/carbon nanotubes and epoxy/cnt composites. *Polym. Plast. Technol. Eng.* **2016**, *55*, 1167–1191. [CrossRef]
3. Ma, P.-C.; Siddiqui, N.A.; Marom, G.; Kim, J.-K. Dispersion and functionalization of carbon nanotubes for polymer-based nanocomposites: A review. *Compos. Part A* **2010**, *41*, 1345–1367. [CrossRef]
4. Gu, J.; Yang, X.; Li, C.; Kou, K. Synthesis of cyanate ester microcapsules via solvent evaporation technique and its application in epoxy resins as a healing agent. *Ind. Eng. Chem. Res.* **2016**, *55*, 10941–10946. [CrossRef]
5. Lei, L.; Shan, J.; Hu, J.; Liu, X.; Zhao, J.; Tong, Z. Co-curing effect of imidazole grafting graphene oxide synthesized by one-pot method to reinforce epoxy nanocomposites. *Compos. Sci. Technol.* **2016**, *128*, 161–168. [CrossRef]
6. Wang, F.; Drzal, L.T.; Qin, Y.; Huang, Z. Enhancement of fracture toughness, mechanical and thermal properties of rubber/epoxy composites by incorporation of graphene nanoplatelets. *Compos. Part A* **2016**, *87*, 10–22. [CrossRef]
7. Halary, J.L. Structure-property relationships in epoxy–amine networks of well-controlled architecture. *High Perform. Polym.* **2000**, *12*, 141–153. [CrossRef]
8. Rana, D.; Sauvant, V.; Halary, J. Molecular analysis of yielding in pure and antiplasticized epoxy–amine thermosets. *J. Mater. Sci.* **2002**, *37*, 5267–5274. [CrossRef]
9. Domun, N.; Hadavinia, H.; Zhang, T.; Sainsbury, T.; Liaghat, G.; Vahid, S. Improving the fracture toughness and the strength of epoxy using nanomaterials–a review of the current status. *Nanoscale* **2015**, *7*, 10294–10329. [CrossRef] [PubMed]
10. Andrews, R.; Weisenberger, M. Carbon nanotube polymer composites. *Curr. Opin. Solid State Mater. Sci.* **2004**, *8*, 31–37. [CrossRef]
11. Gu, J.; Liang, C.; Zhao, X.; Gan, B.; Qiu, H.; Guo, Y.; Yang, X.; Zhang, Q.; Wang, D.-Y. Highly thermally conductive flame-retardant epoxy nanocomposites with reduced ignitability and excellent electrical conductivities. *Compos. Sci. Technol.* **2017**, *139*, 83–89. [CrossRef]
12. Manchado, M.L.; Valentini, L.; Biagiotti, J.; Kenny, J. Thermal and mechanical properties of single-walled carbon nanotubes–polypropylene composites prepared by melt processing. *Carbon* **2005**, *43*, 1499–1505. [CrossRef]

13. Quaresimin, M.; Salviato, M.; Zappalorto, M. Strategies for the assessment of nanocomposite mechanical properties. *Compos. Part B* **2012**, *43*, 2290–2297. [CrossRef]
14. Odegard, G.M.; Gates, T.S.; Nicholson, L.M.; Wise, K.E. Equivalent-continuum modeling of nano-structured materials. *Compos. Sci. Technol.* **2002**, *62*, 1869–1880. [CrossRef]
15. Wan, H.; Delale, F.; Shen, L. Effect of CNT length and cnt-matrix interphase in carbon nanotube (cnt) reinforced composites. *Mech. Res. Commun.* **2005**, *32*, 481–489. [CrossRef]
16. Liu, Y.; Chen, X. Evaluations of the effective material properties of carbon nanotube-based composites using a nanoscale representative volume element. *Mech. Mater.* **2003**, *35*, 69–81. [CrossRef]
17. Chen, X.; Liu, Y. Square representative volume elements for evaluating the effective material properties of carbon nanotube-based composites. *Comput. Mater. Sci.* **2004**, *29*, 1–11. [CrossRef]
18. Fereidoon, A.; Saeedi, E.; Ahmadimoghadam, B. Comparison between different finite element methods for foreseeing the elastic properties of carbon nanotube reinforced epoxy resin composite. In Proceedings of the World Congress on Engineering, London, UK, 2–4 July 2008; Volume 2, pp. 1381–1385.
19. Shokrieh, M.M.; Rafiee, R. Prediction of mechanical properties of an embedded carbon nanotube in polymer matrix based on developing an equivalent long fiber. *Mech. Res. Commun.* **2010**, *37*, 235–240. [CrossRef]
20. Shojaie, M.; Golestanian, H. Effects of interface characteristics on mechanical properties of carbon nanotube reinforced polymer composites. *Mater. Sci. Technol.* **2011**, *27*, 916–922. [CrossRef]
21. Joshi, U.A.; Sharma, S.C.; Harsha, S. Effect of carbon nanotube orientation on the mechanical properties of nanocomposites. *Compos. Part B* **2012**, *43*, 2063–2071. [CrossRef]
22. Deng, L.; Eichhorn, S.J.; Kao, C.-C.; Young, R.J. The effective Young's modulus of carbon nanotubes in composites. *ACS Appl. Mater. Interfaces* **2011**, *3*, 433–440. [CrossRef] [PubMed]
23. Peng, R.; Zhou, H.; Wang, H.; Mishnaevsky, L., Jr. Modeling of nano-reinforced polymer composites: Microstructure effect on Young's modulus. *Comput. Mater. Sci.* **2012**, *60*, 19–31. [CrossRef]
24. Zuberi, M.J.S.; Esat, V. Investigating the mechanical properties of single walled carbon nanotube reinforced epoxy composite through finite element modelling. *Compos. Part B* **2015**, *71*, 1–9. [CrossRef]
25. Subramanian, N.; Rai, A.; Chattopadhyay, A. Atomistically derived cohesive behavior of interphases in carbon fiber reinforced cnt nanocomposites. *Carbon* **2017**, *117*, 55–64. [CrossRef]
26. Zhigilei, L.V.; Volkov, A.N.; Dongare, A.M. Computational study of nanomaterials: From large-scale atomistic simulations to mesoscopic modeling. In *Encyclopedia of Nanotechnology*; Springer: Dordrecht, The Netherlands, 2012; pp. 470–480.
27. Hu, Z.; Arefin, M.R.H.; Yan, X.; Fan, Q.H. Mechanical property characterization of carbon nanotube modified polymeric nanocomposites by computer modeling. *Compos. Part B* **2014**, *56*, 100–108. [CrossRef]
28. Liu, B.; Huang, Y.; Jiang, H.; Qu, S.; Hwang, K. The atomic-scale finite element method. *Comput. Methods Appl. Mech. Eng.* **2004**, *193*, 1849–1864. [CrossRef]
29. Liu, L.; Barber, A.H.; Nuriel, S.; Wagner, H.D. Mechanical properties of functionalized single-walled carbon nanotube/poly (vinyl alcohol) nanocomposites. *Adv. Funct. Mater.* **2005**, *15*, 975–980. [CrossRef]
30. Chen, G.-X.; Kim, H.-S.; Park, B.H.; Yoon, J.-S. Multiwalled carbon nanotubes reinforced nylon 6 composites. *Polymer* **2006**, *47*, 4760–4767. [CrossRef]
31. Shokrieh, M.; Mahdavi, S. Micromechanical model to evaluate the effects of dimensions and interphase region on the elastic modulus of CNT/polymer composites. *Modares Mech. J.* **2011**, *11*, 13–25.
32. Anderson, T.L.; Anderson, T. *Fracture Mechanics: Fundamentals and Applications*; CRC Press, Taylor & Francis Group: Boca Raton, FL, USA, 2005.
33. Bahr, J.L.; Tour, J.M. Covalent chemistry of single-wall carbon nanotubes. *J. Mater. Chem.* **2002**, *12*, 1952–1958. [CrossRef]
34. Gou, J.; Minaie, B.; Wang, B.; Liang, Z.; Zhang, C. Computational and experimental study of interfacial bonding of single-walled nanotube reinforced composites. *Comput. Mater. Sci.* **2004**, *31*, 225–236. [CrossRef]
35. Weidt, D.; Figiel, L. Effect of cnt waviness and van der Waals interaction on the nonlinear compressive behaviour of epoxy/cnt nanocomposites. *Compos. Sci. Technol.* **2015**, *115*, 52–59. [CrossRef]
36. Zakeri, M.; Shayanmehr, M.; Shokrieh, M. Interface modeling of nanotube reinforced nanocomposites by using multi-scale modeling method. *Modares Mech. Eng.* **2013**, *12*, 1–11.
37. Ayatollahi, M.; Shadlou, S.; Shokrieh, M. Multiscale modeling for mechanical properties of carbon nanotube reinforced nanocomposites subjected to different types of loading. *Compos. Struct.* **2011**, *93*, 2250–2259. [CrossRef]

38. Jiang, L.; Huang, Y.; Jiang, H.; Ravichandran, G.; Gao, H.; Hwang, K.; Liu, B. A cohesive law for carbon nanotube/polymer interfaces based on the van der waals force. *J. Mech. Phys. Solids* **2006**, *54*, 2436–2452. [CrossRef]
39. Tan, H.; Jiang, L.; Huang, Y.; Liu, B.; Hwang, K. The effect of van der waals-based interface cohesive law on carbon nanotube-reinforced composite materials. *Compos. Sci. Technol.* **2007**, *67*, 2941–2946. [CrossRef]
40. Park, K.; Paulino, G.H.; Roesler, J.R. A unified potential-based cohesive model of mixed-mode fracture. *J. Mech. Phys. Solids* **2009**, *57*, 891–908. [CrossRef]
41. Spring, D.W.; Paulino, G.H. A growing library of three-dimensional cohesive elements for use in abaqus. *Eng. Fract. Mech.* **2014**, *126*, 190–216. [CrossRef]
42. Spring, D.W.; Paulino, G.H. Computational homogenization of the debonding of particle reinforced composites: The role of interphases in interfaces. *Comput. Mater. Sci.* **2015**, *109*, 209–224. [CrossRef]
43. Born, M.; Huang, K. *Dynamical Theory of Crystal Lattices*; Clarendon Press: Oxford, UK, 1954.
44. Zhang, P.; Huang, Y.; Gao, H.; Hwang, K. Fracture nucleation in single-wall carbon nanotubes under tension: A continuum analysis incorporating interatomic potentials. *J. Appl. Mech.* **2002**, *69*, 454–458. [CrossRef]
45. Glaucio, H. Paulino on the Web. Available online: https://paulino.ce.gatech.edu/PPR_tutorial.html (accessed on 17 June 2017).
46. Kundalwal, S.; Meguid, S. Multiscale modeling of regularly staggered carbon fibers embedded in nano-reinforced composites. *Eur. J. Mech. A-Solids* **2017**, *64*, 69–84. [CrossRef]
47. Alian, A.; Kundalwal, S.; Meguid, S. Interfacial and mechanical properties of epoxy nanocomposites using different multiscale modeling schemes. *Compos. Struct.* **2015**, *131*, 545–555. [CrossRef]
48. Alian, A.; Kundalwal, S.; Meguid, S. Multiscale modeling of carbon nanotube epoxy composites. *Polymer* **2015**, *70*, 149–160. [CrossRef]
49. Tserpes, K.; Chanteli, A. Parametric numerical evaluation of the effective elastic properties of carbon nanotube-reinforced polymers. *Compos. Struct.* **2013**, *99*, 366–374. [CrossRef]
50. US Research Nanomaterials, Inc. Available online: us-nano.com (accessed on 5 May 2017).
51. Shokrieh, M.M.; Rafiee, R. Stochastic multi-scale modeling of CNT/polymer composites. *Comput. Mater. Sci.* **2010**, *50*, 437–446. [CrossRef]
52. Zeng, Q.; Yu, A.; Lu, G. Multiscale modeling and simulation of polymer nanocomposites. *Prog. Polym. Sci.* **2008**, *33*, 191–269. [CrossRef]
53. Li, C.; Chou, T.-W. Multiscale modeling of compressive behavior of carbon nanotube/polymer composites. *Compos. Sci. Technol.* **2006**, *66*, 2409–2414. [CrossRef]
54. Ma, P.-C.; Mo, S.-Y.; Tang, B.-Z.; Kim, J.-K. Dispersion, interfacial interaction and re-agglomeration of functionalized carbon nanotubes in epoxy composites. *Carbon* **2010**, *48*, 1824–1834. [CrossRef]
55. Liu, C.-X.; Choi, J.-W. Improved dispersion of carbon nanotubes in polymers at high concentrations. *Nanomaterials* **2012**, *2*, 329–347. [CrossRef] [PubMed]
56. Bhattacharya, M. Polymer nanocomposites—A comparison between carbon nanotubes, graphene, and clay as nanofillers. *Materials* **2016**, *9*, 262. [CrossRef] [PubMed]

© 2018 by the authors. Licensee MDPI, Basel, Switzerland. This article is an open access article distributed under the terms and conditions of the Creative Commons Attribution (CC BY) license (http://creativecommons.org/licenses/by/4.0/).

nanomaterials

MDPI

Article

Effect of Functionalized Graphene Nanoplatelets on the Delamination-Buckling and Delamination Propagation Resistance of 3D Fiber-Metal Laminates Under Different Loading Rates

Davide De Cicco and Farid Taheri *

Advanced Composites and Mechanics Laboratory, Department of Mechanical Engineering, Dalhousie University, 1360 Barrington Street, P.O. Box 15 000, Halifax, NS B3H 4R2, Canada; davide.decicco@dal.ca
* Correspondence: farid.taheri@dal.ca; Tel.: +1-902-494-3935; Fax: +1-902-484-6635

Received: 10 September 2019; Accepted: 11 October 2019; Published: 18 October 2019

Abstract: This paper presents an investigation into the effect of graphene nanoplatelets (GNPs) as a means of improving the impact buckling performance and delamination propagation resistance of a recently developed 3D fiber-metal laminate (3D-FML). One of the highlights of the investigation is the examination of the performance of the GNP-reinforced resin at a sub-freezing temperature (−50 °C). 3D-FML beam specimens were subjected to axial impact of various intensities at room-temperature, while they were subjected to quasi-static axial compression load at the sub-freezing temperature. Moreover, the influence of two different surface preparation methods on the performance of the metallic/FRP interfaces of the hybrid system was also investigated in this study. Although the inclusion of the GNPs in the resin resulted in some gain in the buckling capacity of the 3D-FML, nevertheless, the results revealed that the lack of adequate chemical bond between the GNP-reinforced resin and the magnesium skins of the hybrid material system significantly limited the potential influence of the GNPs. Therefore, a cost-effective and practical alternative is presented that results in a significant improvement in the interfacial capacity.

Keywords: 3D fiber-metal laminates; graphene nanoplatelets; impact buckling; delamination buckling; delamination propagation; temperature effect

1. Introduction

The automobile industry, like many other industries, is facing challenges in complying with the recent and continually increasing strict environmental regulations and safety requirements. Therefore, the development of cost-effective performant materials, ranging from light-weight metallic alloys to various composites, are being increasingly explored to improve vehicles' fuel consumption and passenger safety. In response, new metallic alloys' compositions are becoming increasingly complex in order to optimize their performance (cf. [1,2]). However, one of the proven strategies for developing light-weight hybrid materials has been shown to be attained by the marriage of lightweight metal alloys and advanced fiber-reinforced polymers (FRPs), referred to as fiber-metal laminates (FMLs). Following this path, our research group recently developed a new class of FML, consisting of a truly three-dimensional hollow-core fiberglass fabric, with its core filled with a light-weight foam, sandwiched between thin sheets of a lightweight magnesium alloy. This hybrid composite system, shown in Figure 1, is referred to as 3D-FML. Due to the impressive specific strength, stiffness and impact absorption properties of this hybrid system, 3D-FMLs are considered as economical and effective light-weight material systems, suitable for the fabrication of transport vehicles and aircraft shell structures [3,4]. However, as highlighted in some of the authors' previous works [5–7], the outstanding performance of this class of FML is somewhat compromised when the system becomes subject to in-plane loading. This is mainly due to the relatively low strength of the magnesium/FRP interface segment of this FML. In other words, in general, the interface strength between the metallic and FRP constituents is the Achilles' heel of essentially all classes of FMLs. Therefore, various methods and strategies have been explored to increase the interfacial bond strength; these approaches can be essentially divided into the following categories: (i) abrasion [8–10], (ii) chemical etching (which includes plasma surface activation) [11–14], (iii) use of nanoparticles to reinforce the interface resin, and (iv) a combination of the aforementioned methods.

Abrasion enhances the mechanical bond (interlocking) between the substrate and the resin, while chemical etching improves the chemical bond between the two interfacing materials. The use of nanoparticles (NPs), however, strengthen the resin used for bonding the two substrates and bridges the potential micro-cracks and their growth, as well as enhancing the interlocking between the substrates' micro-cavities and the adhesive. It is worth noting that NPs themselves cannot enhance the chemical bond between the adhesive and substrate. Therefore, in the absence of an effective chemical bond, the effectiveness of mechanical interlocking becomes significantly compromised [15].

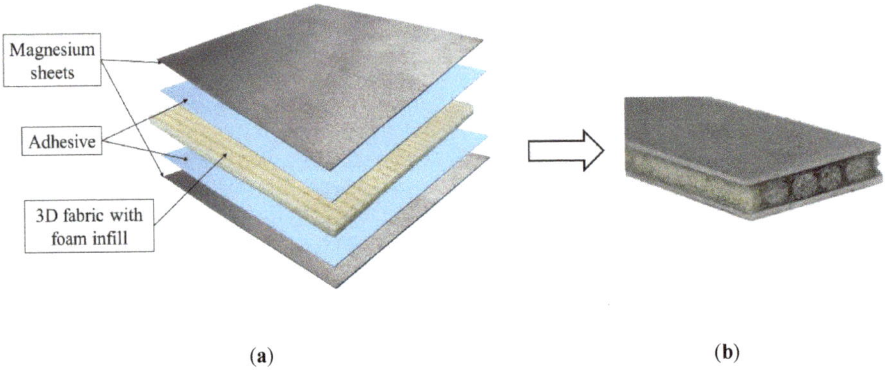

Figure 1. The 3D-FML composite: (**a**) exploded view, showing the various components, and (**b**) the final product.

In this paper, the influence of incorporation of graphene nanoplatelets (GNPs) on enhancing the interfacial bond strength is investigated. It should be noted that other effective types of nanoparticles are also available for the purpose (see for instance the use of nanosilica [16–18], nanoclay [19–21], polymers themselves [22,23], and, for specific medical applications, nanogold and nanosilver, [24–26]). Several studies have shown that the modulus of elasticity, tensile strength, and fracture toughness [27–34], as well as fatigue resistance [35,36] and vibration damping capacity [37–40] of polymers could be positively enhanced by the incorporation of appropriate NPs. NPs have also been shown to improve the bond strength of adhesively bonded joints, especially in lap-strap joints [27,41,42].

Improvements in impact resilience gained by the use of NPs have also been demonstrated. Haro et al. [43] and Áliva et al. [44] performed ballistic impact tests on a Kevlar/aluminum FML and fiberglass/epoxy composite, respectively. They observed that alumina, silica and nanoclay NP-reinforcements, respectively, led to increasing ballistic protection; however, the inclusion of the NPs caused additional delamination extension. Haq et al. [45] performed low-velocity impact tests on sandwich composites. They reinforced the fiberglass/epoxy facial laminates by coating them with graphene. The authors concluded that an optimal spatial distribution of NPs could be done to optimize the response of components subjected to impact. Some researchers have also shown that GNPs were more effective than carbon nanotubes (CNTs) in delamination mitigation and in arresting crack propagation [46,47]. For instance, Rafiee et al. [48] demonstrated that while the incorporation of CNTs in epoxy led to a 20% increase in the fracture toughness, a 53% increase was obtained when GNPs was used as the reinforcement. In addition, Ahmadi-Moghadam et al. [49,50] demonstrated that further improvements in fracture toughness (in all three modes) could be attained through functionalization of GNPs. Functionalization is a chemical process aimed at improving the interfacial bond between NPs and resins. The authors showed that amongst the four different functionalization schemes they tried, the best results were obtained when NH_2 was used to functionalize the GNPs.

There are, however, studies that report the incorporation of NPs did not lead to beneficial outcomes. For instance, Wichman et al. [51] observed some increase in the mechanical properties of their CNT-reinforced fiberglass/epoxy, however, no improvement in the delamination resistance was attained. Using the same type of NPs, Siegfried at al. [52] reported an increase in impact performance of carbon fiber/epoxy composites subjected to low-velocity impact, but at the expense of increased delamination. The authors attributed the loss of the interlaminar strength to an increase in the matrix brittleness due to the incorporation of NPs. Bortz et al. [53] reported an increase in the stress intensity factor of 63% when GNPs were incorporated into the resin, thus, showing how the composite was more prone to cracking.

Since this paper focuses on the compressive axial behaviour of 3D-FML, it is worth mentioning some of the notable works related to the study of buckling in composites, more specifically, related to composites that have a delamination. The presence of a delamination in composites has been shown to negatively affect their response, especially when the composite is subjected to an in-plane compressive loading [54–58]. Delamination can be initiated in FRP due to even a low-energy impact (i.e., caused by the impact of a falling tool during fabrication) and/or other manufacturing induced flaws. Esfahani et al. [59] carried out a numerical study and showed that the presence of a delamination had a negative impact on the buckling capacity of their specimens, especially when the delamination was close to the specimen's outer laminae. Kim and Hong [60] reported that the length and position of the delamination were two parameters that had a large influence on the buckling mode and post-buckling behaviour of laminated composites. It should, however, be noted that there exists a threshold under which the delamination length would not affect the buckling strength of composites. Asaee et al. [61] demonstrated the efficacity of using GNPs in improving the in-plane static compression response of short 3D-FML beams. By adding the GNPs to the bonding interface between the magnesium and core part of their hybrid material system, they observed up to 25% increase in the specimens' load-bearing capacity.

When considering transport vehicles, in addition to the conventional mechanical loads (including impact loads), they also become subjected to severe environmental conditions, including extreme temperatures and humidity. In many areas in the world, temperatures as low as −50 °C are commonly reached and maintained during the winter period. Therefore, it is important to assure the durability of materials used in fabricating transport vehicle panels, especially in circumstances when the vehicle becomes subjected to an impact within the harsh environment. Several studies have investigated the effect of temperature on the performance of composite materials. For example, Taraghi et al. [62] subjected Kevlar/epoxy composite to low-velocity impacts at room and sub-freezing (−40 °C) temperatures and observed 35% and 34% reduction in damage density, respectively, as a result of the inclusion of CNTs to their epoxy resin. Shen at al. [63] showed that the inclusion of graphene oxide particles improved the interlaminar shear strength of fiberglass/epoxy composites by 32% at a cryogenic temperature of 77 K (−196 °C). A few authors have also considered the influence of thermal cycles on materials [64,65]. For instance, Khalili et al. [66] investigated the influence of moisture and the subsequent thermal cycles on the strength of hybrid bonded/bolted joints mating FML substrates. The specimens were initially soaked in seawater for 30 days and were subsequently subjected to 10 thermal cycles (between −40 °C to −100 °C). They observed a 35% reduction in the strength of their immersed specimens. However, the cooling cycles recuperated 50% of the lost strength in their joints. This gain in the strength was believed to have occurred as a result of the relaxation of the residual stresses developed in the immersed specimens.

In this paper, the in-plane compression and impact-buckling responses of 3D-FML whose metal/FRP interfaces are reinforced with NH_2-functionalized GNPs are investigated. In particular, the influence of the existence of a delamination in the FML is also considered. The responses of reinforced specimens are compared against the baseline specimens (i.e., with non-reinforced specimens). Moreover, the influence of sub-freezing temperature (−50 °C) on the performance of the non-reinforced and GNP-reinforced interfaces subjected to quasi-static compression loading is also investigated.

2. Materials and Methods

2.1. Materials

The 3D fiberglass fabric and fiberglass veil were acquired from China Beihai Fiberglass Co. Ltd. (Jiujiang City, Jiangxi, China). A Huntsman produced two-part hot-cure epoxy resin (bisphenol-A-based Araldite LY1564 resin and its Aradur 2954 hardener) was acquired from the producer (Huntsman Corporation, West Point, GA, USA), while the cold-cure epoxy resin (105 resin with 206 hardener) used to mate the magnesium and FRP constituents (i.e., the interface region herein) was produced by West System (Bay City, MI, USA). An 8-lb/ft^3, high-density polyurethane foam was obtained from US Composites (West Palm Beach, FL, USA). The NH_2 functionalized graphene nanoplatelets (hereafter referred to as GNPs for the sake of brevity), having an in-plane dimension of 1–2 μm and thickness of 4 nm, were purchased from CheapTubes Inc. (Cambridgeport, VT, USA). The lightweight AZ31B-H24 magnesium alloy sheets (or skins) were acquired from MetalMart (Commerce, CA, USA). Finally, liquid nitrogen was obtained locally.

2.2. Specimen's Fabrication

All the beam-like specimens, schematized in Figure 2, with dimensions of 190 mm × 20 mm × 5.3 mm, were extracted from larger 3D-FML plates, using a water-cooled circular saw equipped with a diamond-coated blade. The sequence of procedures used to fabricate the plates is as follows. The two parts of the hot-cure epoxy resin were mixed at 100 rpm for 10 min. using a mixer, then degassed in a vacuum chamber for a minimum of 30 min. Then, the mix was applied homogeneously onto the 4 mm thick 3D fiberglass fabric (3D-FGF) using a brush. The resin-immersed fabric was cured at 60 °C for two hours and subsequently at 120 °C for 8 h, after which the fabric took its three-dimensional configuration with cavities in its core (see Figure 1). The cavities were then filled with the foam to provide support to the thin pillars connecting the two main biaxial E-glass constituents of the fabric, thereby increasing the overall specimen's stiffness and strength. The foam-filling process was done by drawing the foam into the cavities at its liquid stage under a negative pressure using an in-house designed jig, which guarantees a homogeneous repartition of the foam inside the cavities. The combination of the 3D fabric-epoxy and foam will be referred to as "core" hereafter.

The hybrid sandwich composite system was completed by bonding the magnesium plates (skins) to the core. Two bonding methods were used, thus leading to two different categories of specimens. In both methods, first, the magnesium skins' bonding surfaces were sandblasted with coarse 20–30 grit crushed glass abrasive in order to facilitate good mechanical bonding. The two bonding methods are as follows:

(i) in the first method (referred to as SB, hereafter), the hot-cure resin was directly applied onto the substrates (skins and core), and then the resulting sandwich was vacuum bagged and cured for two hours at 60 °C and eight hours at 120 °C.
(ii) in the second method (referred to as SBC, hereafter), the magnesium bonding surfaces were pre-coated with a thin layer of cold-cure resin, cured for 24 h under vacuum. In a second step, another layer of cold-cure resin was applied to both adherends and they were sealed under vacuum and let cure at room temperature for 24 h. This second method was developed by the authors and the resulting gain in the interface bond strength under different loading conditions, including axial impact loading, was reported in [67].

For the specimens hosting a delamination, a thin sheet of Teflon was inserted between the magnesium and the core during the bonding process. The resulting delamination had a length of 30%, 50% or 70% of the specimens' gage length, and were placed at mid-length, on one of the interfaces only (see Figure 2).

In some specimens, GNPs were incorporated into the resin that was used to adhere the skins to the core. This was done according to the detailed procedure described in [42]. In brief, the various weight percentages (wt%) of the functionalized-GNPs nanoparticles were accurately measured using a scale having a precision of 0.5 mg. The GNPs were mixed with the cold-cure resin (part 105) using a variable speed mixer for 15 min. with an initial speed of 400 rpm, gradually increasing to 2000 rpm. Then, the mixture was further homogenized by passing it seven times through a three-roll calendering machine to break the agglomerations and facilitate uniform dispersion of the particles. Finally, the hardener (part 206) was incorporated, and the whole mixture was mixed at 400 rpm for four minutes and subsequently degassed for five minutes. The short mixing and degassing times prevent the resin from partially curing before it is applied onto all the adherends' surfaces. After degassing, the resin was used as an adhesive in the same manner as described earlier.

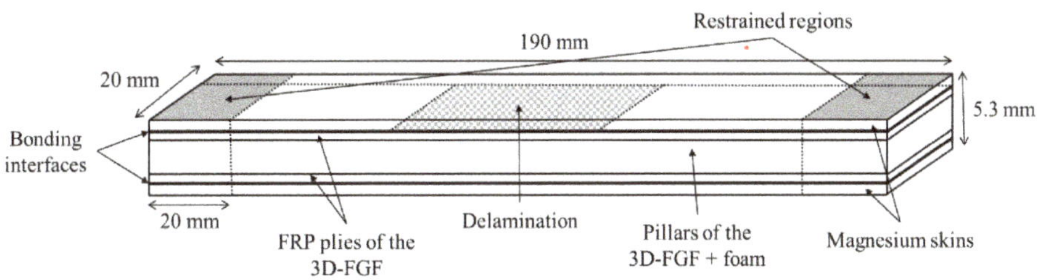

Figure 2. Schematic illustration of the 3D-FML specimen hosting a delamination and its overall dimensions (drawing not to scale).

2.3. *Testing Apparatus, Procedures and Data Acquisition*

2.3.1. Case Studies I

The experimental investigation of this study was organized within three distinct case studies (I, II, and III), as summarized in Figure 3.

In the first study, the integrity of the SB bonding method and the effect of GNP inclusion on the performance of the interface bond was studied. Specimens used in this category were fabricated with four different GNP contents (i.e., no GNPs (referred to as "neat" and identified by "N"), 0.5 wt%, 1 wt% and 2 wt% contents). Each specimen category was subjected to four impact energies (1.5 J, 3 J, 4.5 J, and 7 J). The impact energies were chosen according to an experimental investigation conducted earlier by the authors [6] and are aimed to cause (i) elastic buckling; (ii) initiation of a permanent deformation; (iii) propagation of the delamination and (iv) complete failure of the specimens, respectively.

Four initial delamination scenarios were considered for the neat specimen group; they were: intact (i.e., with no initial delamination), identified as ND; and those with three different initial delamination lengths equal to 30%, 50% and 70% of the specimens' gage length. For the specimens that contained GNPs, only the intact specimens and the specimens with initial delamination length of 50% were considered. Detailed justification of the selection criteria is provided in the next sections.

It should be noted that the effect of the inclusion of GNP on the interface bond strength (i.e., case study I) was performed previously. However, as will be explained in detail in the subsequent section, the benefits that could be gained by the inclusion of GNPs in the resin were rendered inconclusive. Therefore, the new bonding procedure described in the previous section was used to form the case studies II and III.

2.3.2. Case Study II

For the second case study, one delamination length (i.e., 50% of the gage length), one impact energy (2.85 J), and one GNP content (0.5 wt%) were considered. In this way, more statistical number of specimens could be considered per testing category. The 2.8 J impact energy was selected on the basis that it would cause partial buckling of the specimens and propagating the initial delamination while preventing the complete failure of the specimens. Note that the complete delamination of the skin would defeat one of the objectives of the study (i.e., the examination of GNP's effect on the delamination growth). Furthermore, the inclusion of GNPs in the resins were done in two ways:

(i) included only in the resin used to coat the magnesium skins was reinforced with the GNPs (these specimens are identified as "C" specimens);
(ii) included in the resin used to coat the skins and in the resin used for bonding the skins to FRP were both reinforced with the GNPs (specimens of this category are identified by "CA").

Aside from the GNPs, the effect of inserting a thin fiberglass veil between the magnesium and the core with the aim of improving the interface bond mechanism was also investigated (specimens in this category are identified as "V" specimens). Finally, the baseline specimens, which were fabricated with the neat resin (i.e., with no GNP or veil reinforcements) are identified as "N" specimens.

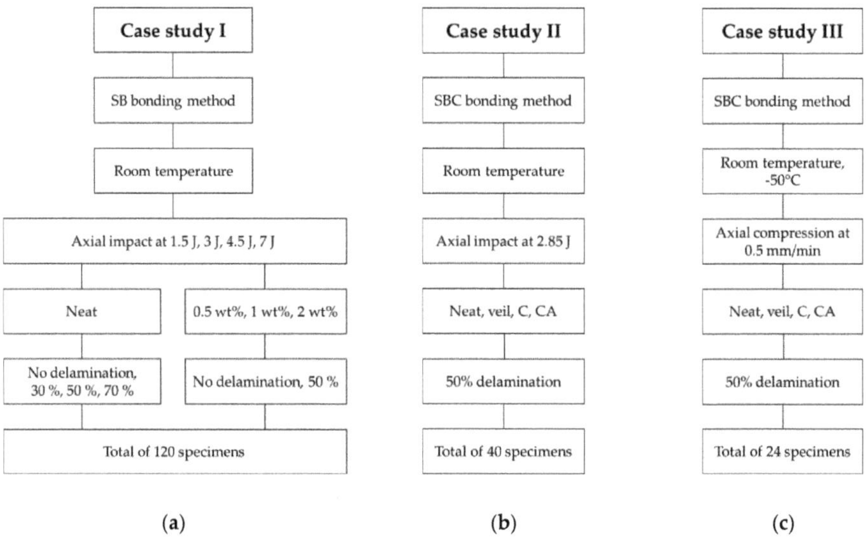

Figure 3. Summary of the case studies and their parameters: (**a**) I, (**b**) II, and (**c**) III.

2.3.3. Impact Testing Apparatus

The impact testing apparatus used to test the specimens of case studies I and II is shown in Figure 4. The setup consisted of a modified Charpy impact testing machine equipped with an in-house designed fixture to support the specimen such that a given specimen would be subjected to a purely axial impact. Each specimen was clamped in the fixture over 20 mm length at each end in such a way that only the axial displacement at the impacted end was permitted, therefore, imposing a fixed-fixed boundary condition. The various impact energies were obtained through trial and error, by changing the pendulum angle and using an image-processing algorithm written in MATLAB, to extract the position and time information of the impactor. The impact load and axial-shortening history data were captured using a dynamic load cell and a dynamic linear variable displacement transducer (DLVDT), respectively, both operated at a sampling rate of 50 kHz. The signals were transferred to a PC via a

National Instruments data acquisition system device, synchronized using the Signal Express software. A Photron Fastcam PCI high-speed camera was used to record the impactor movement and specimens' deformation at a rate of 2000 fps for the first case study, while a Kronotek Chronos high-speed camera was used for the case study II tests, at a rate of 4498 fps. Note that the latter camera was not available to the authors at the time the first case study was conducted (hence, the use of two different cameras).

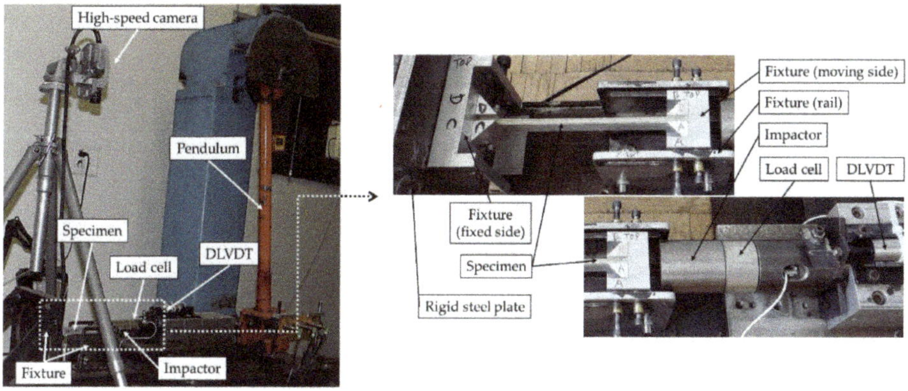

(a) (b)

Figure 4. (**a**) Experimental setup for impact testing and (**b**) close-up view of the impactor and the specimen supported by the fixture.

2.3.4. Case Study III

The same parameters that were used in case study II were used in case study III, but the tests were conducted under quasi-static loading (cf. Figure 3). This is because the sub-freezing temperature had to be conducted in an Instron thermal chamber that could not accommodate the impact test setup.

The chamber was used in conjunction with an MTS servo-hydraulic testing machine, equipped with a 250 kN load cell. This test setup is illustrated in Figure 5. The compression actuation speed was set to 0.5 mm/min. The air inside the chamber was cooled down to −50 °C using liquid nitrogen, and the temperature was monitored using a thermocouple. Finally, the load and displacement data were retrieved directly from the MTS machine using the MTS793 software that was used to control it, while the delamination-buckling event was captured on video at a rate of 30 fps using a Rebel SL2 camera (Canon, Tokyo, Japan).

2.4. Data Processing

A LabVIEW algorithm was developed to facilitate the post-processing of the impact test data in a consistent and efficient manner. The only required operation of the user was the identification of the exact initial time of the impact event. The output of the LabVIEW code was a set of data points corresponding to three signals. First, the captured displacement-time signal was filtered to remove high-frequency noise and the 60 Hz noise originated by the power supply, followed by filtering of the force-time signal. An example of such signals is illustrated in Figure 6a. Moreover, since the inherent signal fluctuation makes it difficult to objectively compare the signals obtained from testing various specimens, therefore, the RMS (acronym of the root-mean-square) of the signals was obtained, as illustrated in Figure 6b. The RMS data was established by evaluating the average of the load-history signal, computed using the RMS amplitude of the signal. This quantity is directly proportional to the signal's power and peak amplitude. Therefore, the information conveyed through the RMS signal is equivalent to the one from the filtered signal from which they are extracted. The application of this signal-processing procedure would not be necessary when analyzing the quasi-static test results, since there would be no such inherent fluctuation in the signals in such tests.

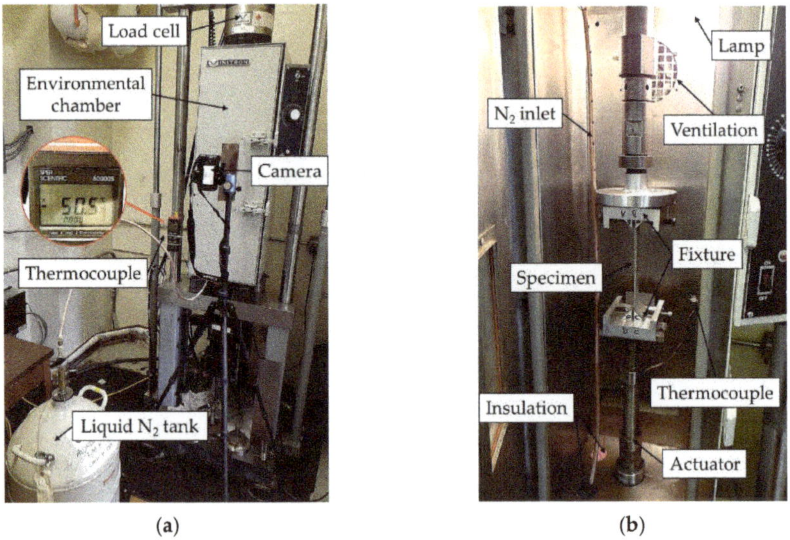

Figure 5. (**a**) Overall view of the static buckling test setup and (**b**) inside view of the thermal chamber.

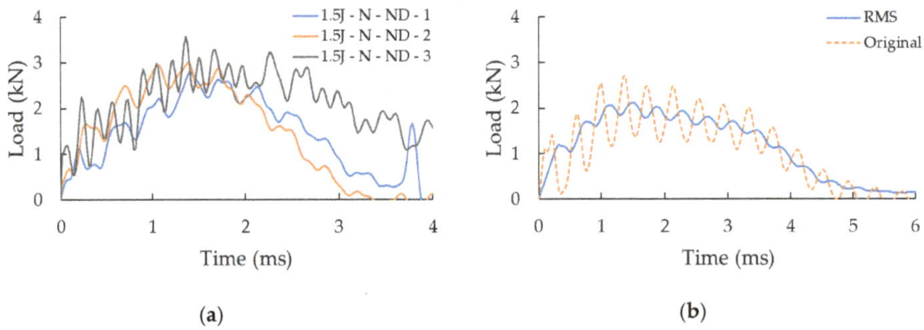

Figure 6. (**a**) The raw load history signals of neat specimens (i.e., without al delamination) impacted at 1.5 J, (**b**) A typical filtered signal and its RMS.

To measure the delamination growth, the initial delamination was precisely measured using a digital microscope and its extremities were marked using a permanent marker. Then, clearly visible tick marks, spaced at 5 mm intervals, were inscribed along the specimens' side. The delamination growth in each specimen was then measured by comparing the images (extracted from the videos) of the specimen at its initial and deformed states.

3. Results and Discussion

In this section, the results of the three case studies are reported and discussed. For the sake of clarity and brevity, only the response of a typical specimen per group of specimens will be illustrated, with the proviso that the exhibited curves are close representatives of the response of all specimens tested within each specimen group. The level of consistency in typical date is illustrated in Figure 6a.

3.1. Case Study I

Typical qualitative responses of an intact (neat) and a specimen having a delamination are shown in Figure 7. In general, the specimens remained straight during the first instance upon the application of the impact load, regardless of the considered impact energies. The intact specimens subsequently experienced global buckling. The specimens that were subjected to 1.5 J impact energy endured the energy by elastic deformation and fully recovered their original status after the event. Those undergoing 3 J impact energy, also underwent global buckling, however, ending up with a permanent deformation since their magnesium skins endured some degree of plastic deformation. The behaviours of the specimens undergoing 4.5 J impact energy was similar to those that were subjected to 3 J impact energy, with the difference that one of the skins partially delaminates in this category. Finally, for 7 J impact energy case, the specimens' skin, on the side that underwent compression during the buckling event delaminated, and the FRP plies of the 3D-FGF on the compression side crushed, leading to the complete failure of the specimens.

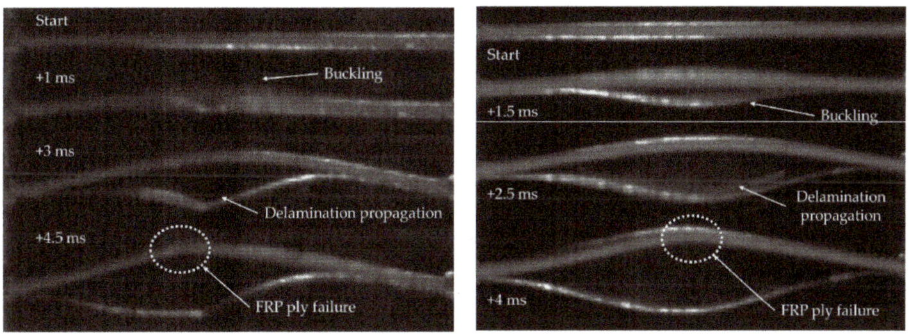

(a) (b)

Figure 7. Illustration of the behaviour of the 3D-FML sandwich under axial impact, for the study case I neat specimens. (**a**) No initial delamination, 7 J and (**b**) 50% initial delamination, 4.5 J.

The specimens having a delamination experienced a global buckling mode; however, during the buckling, the delaminated portion of the skin also experienced local buckling. The delamination then grew to a certain extent depending on the applied impact energy. The propagation of delamination was observed to be marginal in the specimens that were subjected to the lowest impact energy. However, the delamination propagated along the entire span of the specimens that were subjected to 3 J impact, but their core remained undamaged. Finally, the specimens that experienced 4.5 J energy failed completely (i.e., in addition to complete separation of their skins, their FRP plies also failed in compression).

The influence of the presence of a delamination is presented quantitatively in Figure 8 through the load-history graphs of the impact tests performed at 3 J and 7 J on the neat specimens (i.e., specimens without GNPs added to their interfaces). Note that the results of the tests conducted at other two energy levels were omitted for the sake of conciseness since they followed the same pattern. The graphs show a clear reduction of the buckling capacity (corresponding to the maximum load on the curves) for the specimens hosting a delamination compared to the intact specimens. More specifically, reductions in buckling capacity of 26%, 36%, 38%, and 24%, respectively, are observed for specimens experiencing the impact energies of 1.5 J, 3 J, 4.5 J and 7 J. As evident in the curves illustrated in Figure 8, the length of the delamination does not seem to affect the response in a significant manner; in other words, the variation in the impact response is negligible in all three delamination lengths. Consequently, the 30% and 70% delamination cases were not considered for the remaining portion of the study.

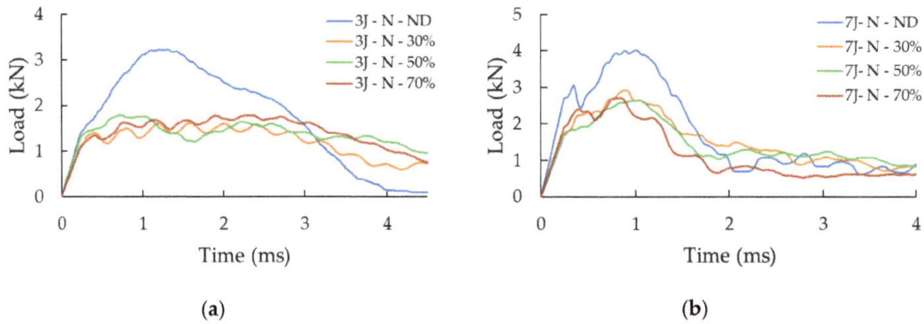

Figure 8. Effect of initial delamination for (**a**) 3 J and (**b**) 7 J cases.

Figure 9 illustrates the load-history graphs for specimens that were subjected to the four impact energies. As could be expected, a higher impact energy led to a higher measured maximum load-bearing capacity. Overall, the results are more consistent for the lowest and highest energies than for the two medium energies. As discussed in [6], this is attributed to the fact that 3 J and 4.5 J energies hover around the energy that corresponds to the damage threshold. Therefore, the sensitivity to the reaction of a given specimen at the onset of buckling, which is naturally volatile, is further amplified. It can also be seen that the specimens tested at the two higher impact energies appear to exhibit a residual load-bearing capacity. This response is not observed when considering the specimens of the other cases because, in those cases, the load drops to zero when the impactor detaches from the specimen (bounces back). In the 4.5 J impact event, the impactor speed halts to zero but without bouncing, indicating that the impact energy is absorbed fully by the specimens, while under 7 J energy, the specimens are completely crushed by the impact. This shows that once the skins are fully delaminated, the strength of the core is fully compromised.

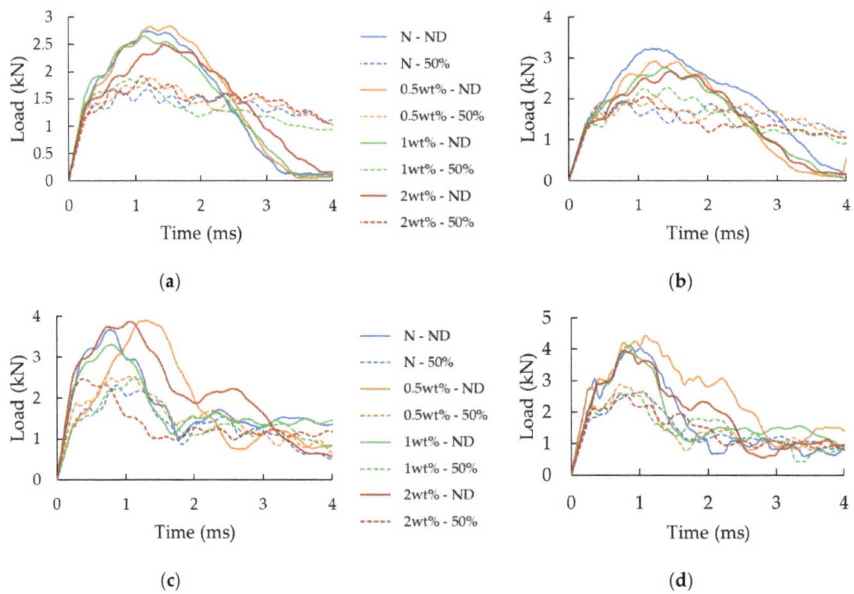

Figure 9. Effect of nanoparticles for specimens with and without delamination, and with and without 0.5 wt% GNP content subjected to the different impact energies: (**a**) 1.5 J, (**b**) 3 J, (**c**) 4.5 J, and (**d**) 7 J.

The effects of the GNPs inclusion can also be observed from the results reported in Figure 9, with a more concise comparison illustrated in Figure 10. Note that all the results shown in Figure 10 have been normalized with respect to the performance of the intact neat specimens, which are referred to as the "baseline" specimens hereafter. The standard deviations are also reported in the chart to better quantify the variation in the results.

To provide the reader with a more comprehensive sense of variation is the results, which are somewhat voluminous due to the large number of parameters that were considered, the results are summarized in terms of buckling capacity and reported in Figure 10. In this figure, the results are categorized in numbered boxes for easier comparison. Please note that the capacity is normalized with respect to the baseline specimens (i.e., the specimens without GNPs and without initial delamination, cf. box 1). Box 2 illustrates the results for the specimens that hosted a delamination. As can be seen, there is no distinct difference in the specimens' response as a function of the initial delamination length. This is attributed to the low bonding strength between the magnesium skins and the composite core, as discussed in a previous study [5].

Within the GNP weight contents considered (cf. box 3), the intact specimens with 0.5 wt% of GNP content show the best overall improved performance under all tried energies, followed by those containing 1 wt% and 2 wt% GNP, respectively. More precisely, we can see that the highest gain (i.e., 12.5% increase in load-bearing capacity) is seen in the specimen that was reinforced with 0.5 wt% GNP, tested under 7 J impact energy. Next in the ranking are the specimens that were reinforced by 1 wt% and 2 wt% GNP contents, exhibiting 10.5% gain in load-bearing capacity. A similar conclusion can be drawn for the specimens that were subjected to 1.5 J impact, but the gains are observed only for the specimen that had 0.5 wt% GNP content. In fact, the nanoparticles seem to have induced a negative effect on the specimens that were subjected to the 3 J and 4.5 J cases, since the specimens' load-bearing capacity was reduced.

When the influence of initial delamination is considered (cf. box 4), the best results are still shown by the specimens that were reinforced with 0.5 wt% of GNPs, followed by those with 1 wt% GNP content. The specimens containing 2 wt% GNP did not exhibit any gain in their strength. Note that for the case of 3 J, the specimens with 0.5 wt% GNP exhibited good performance, notwithstanding the fact that the outcome is marginally different when compared to the outcomes associated with specimens containing 1 wt% GNP. Similar to the results observed in the case of the intact specimens, GNP inclusion resulted in a detrimental effect when the specimens were subjected to 4.5 J impact energy; however, improvement in performance are also observed in the cases when the applied impact energies were 1.5 J and 3 J. Note that the specimens containing 2 wt% GNP content that were subjected to the highest energy performed the least favourably.

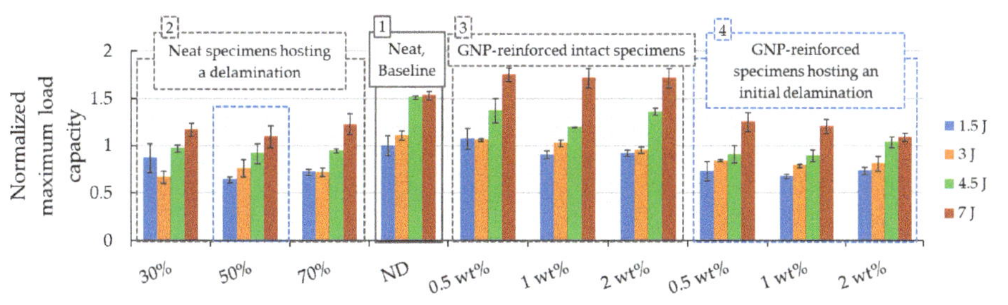

Figure 10. Average normalized buckling capacity for all specimen sets of case study I. Note that the blue boxes identify the comparable specimen categories.

Based on the results, it can be concluded that the addition of GNPs can have beneficial effects on the impact load-bearing capacity of the 3D-FMLs so long as the system has no initial delamination. However, once a delamination is introduced, the lack of an adequate bonding mechanism between the magnesium skin and the resin does not allow the GNPs to play their supportive role in preventing crack initiation and arrest. In the presence of a delamination, the lack of chemical synergy between the magnesium alloy and epoxy resin leads to the catastrophic failure of the interface in the presence of a large magnitude of fracture energy developed by increased loading.

In an attempt to better understand the reason for unanticipated effectiveness of the NPs in suppressing the interface delamination growth of the specimens, the interface bonding surfaces were examined by the use of a digital microscope. The morphology of the surfaces of the specimens within this case study, valid for all the impact energies resulting in delamination, are shown in the micrographs illustrated in Figure 11. One can see that the dispersion of the nanoparticles is homogeneous on the surfaces of the specimens. Note that the darker pixels represent GNPs' distribution and the lighter colour regions seen at the lower portion of each picture correspond to the imprint left by the Teflon that was used to generate the initial delamination. Furthermore, the micrographs in Figure 11e–h illustrate darker magnesium bonding surfaces, which are believed to have occurred as a result of the chemical reaction initiated by the elevated exothermic temperature generated as a result of adhesive's curing process. Furthermore, voids are visible in the adhesive of the specimens that were prepared by the SB bonding method, even with the incorporation of nanoparticles.

Overall, the failure can be classified as the interfacial type. This would suggest that under the present circumstances, one could gain only a marginal enhancement in the interfacial strength as a result of the inclusion of NPs within the interface, unless one could generate a stronger bond between the epoxy adhesive and magnesium substrate, as a result of which the failure mode could be changed into the desirable cohesive failure.

3.2. Case Study II

Further insight into the effect of inclusion of GNPs on the mitigation of delamination propagation is gained by reviewing the results of the second case study. The behaviour of the specimens during a typical impact event in shown in Figure 12. Similar to the response of the specimens of the first case study, the specimens remained straight for the first compression phase of the loading, followed by the buckling of the delaminated portion of the skin, which initiated the subsequent delamination propagation stage of the event. Note that the delamination propagated in an unstable manner in specimens that were subjected to an in-plane impact loading. In other words, the delamination remained in its initial state as the specimen experienced the load which increased its curvature up to a certain stage of the event. At that stage, however, the critical stress was reached, causing a sudden incremental elongation of the delamination within the specimens after which the equilibrium was regained, leading to stabilization of the load-end shortening response. Finally, the maximum delamination length was attained, at which stage the entire impact energy was consumed by the specimen, and the impactor bounced back.

A comparison of the delamination growth in the tested specimens is illustrated in Figure 13a. The values have been normalized with respect to the average delamination propagation observed in the neat specimens. The delamination is seen to increase with respect to the GNP content, with the worst-case observed when the nanoparticles were added to both the magnesium coating and the resin used to bond the skins to FRP (CA specimens). In those specimens, the final delamination length was twice the length developed in the neat-resin specimens. The best results were achieved when the interface had the fiberglass veil incorporated within. The delamination propagation mitigated in those specimens by an average of 46% when compared to the neat specimens. However, overall, the results exhibit large standard deviations. This is a consequence of the inherently unstable nature of delamination propagation in such brittle mediums. It is worth noting that the standard deviation associated with the specimens that had veiled interface, though relatively large, is the lowest amongst the specimen groups, revealing the slight stabilization of the delamination propagation in those specimens. The observed increase in delamination in specimens containing NPs also corroborates with the observation reported in [52]. Siegfried et al. [52] noted the inclusion of their CNTs led to an increased level of matrix-cracking. This validates our hypothesis that the delamination extends more as the GNP content is increased.

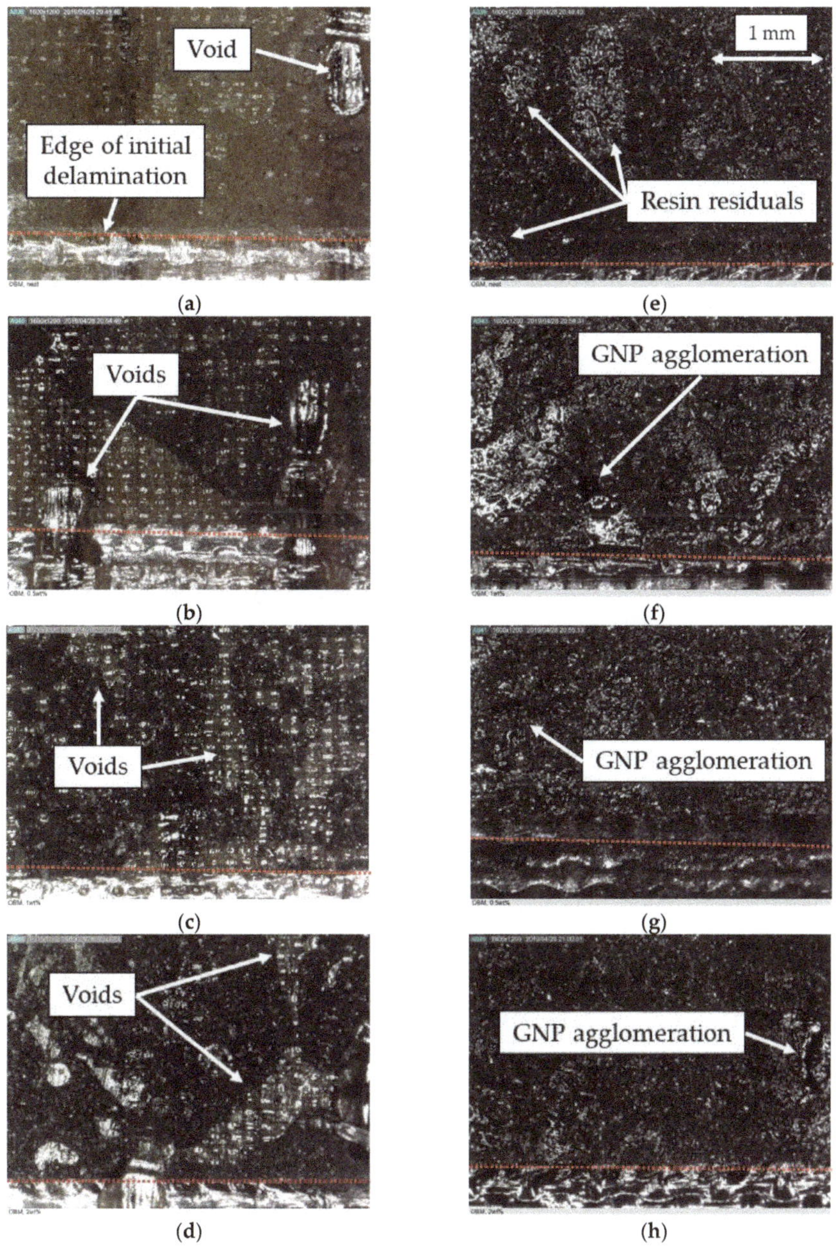

Figure 11. Magnified views of the adherends' fracture surfaces for specimens of case study I. Images (**a**) to (**d**) show the adhesive interfacial surfaces, while images (**e**) to (**h**) show the magnesium interfacial surfaces. From left to right: neat specimens and specimens with 0.5 wt%, 1 wt%, and 2 wt% GNP contents.

The load-bearing capacity of the specimens of this group is reported in Figure 13b. The incorporation of the fiberglass veil into the interface seems to positively impact the load-bearing capacity by increasing it by 6% and reducing the overall standard deviation of the data. In contrast, when GNPs are included only in the epoxy coating (case C), a marginal improvement of 1% is gained. When the standard deviation values are considered, this 1% enhancement in the capacity cannot be considered as a conclusive measure of improvement. Furthermore, the decrease in load-bearing capacity is more pronounced in the specimens of group CA, corresponding to an 8% reduction.

Overall, one could see that the greatest improvement is observed in the specimens that had a fiberglass veil incorporated between their magnesium skins and FRP core. This procedure led to a significant reduction of delamination propagation by 46% and an increase in load-bearing capacity by 6%. Note that the amount of the required effort in implementing the veil is negligible compared to that consumed by the procedure of dispersing the nanoparticles into the resin, which requires mixing and calendaring efforts. Therefore, it can be appreciated that the suggested veil incorporation technique is the more cost-effective alternative.

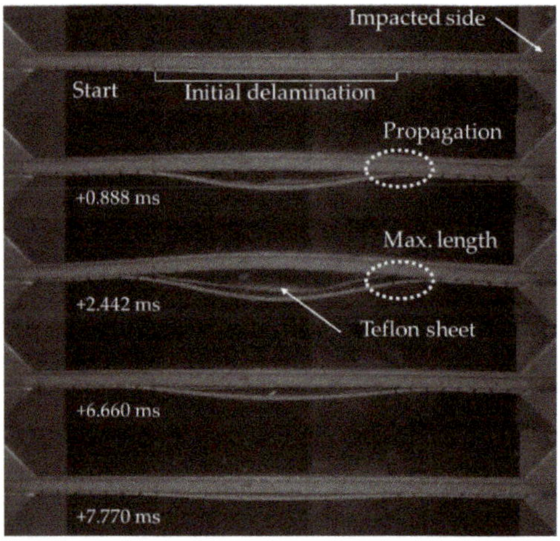

Figure 12. The progressive response of the 3D-FML sandwich under axial impact, for the neat specimens-case study II.

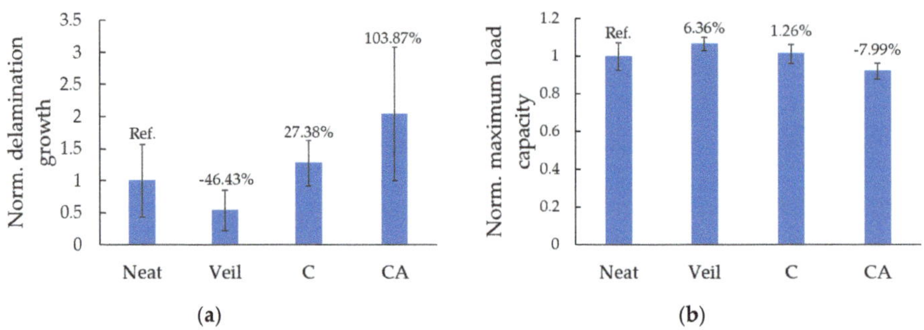

Figure 13. Normalized (**a**) delamination-growth and (**b**) load-bearing capacities for the specimens of case study II (normalized with respect to the "neat" group of specimens).

Similarly to what was done for case study I, micrographs of the bonding surfaces for case study II are provided in Figure 14. Compared to the previous study case, fewer voids are visible at the interface of the specimens that were prepared by the SBC bonding method. In fact, in the latter case, the voids seem to exist only in the specimens that were assembled with the neat adhesive. These observations indicate that the use of nanoparticles and the veil resulted in a more homogeneous distribution of the resin during the curing process.

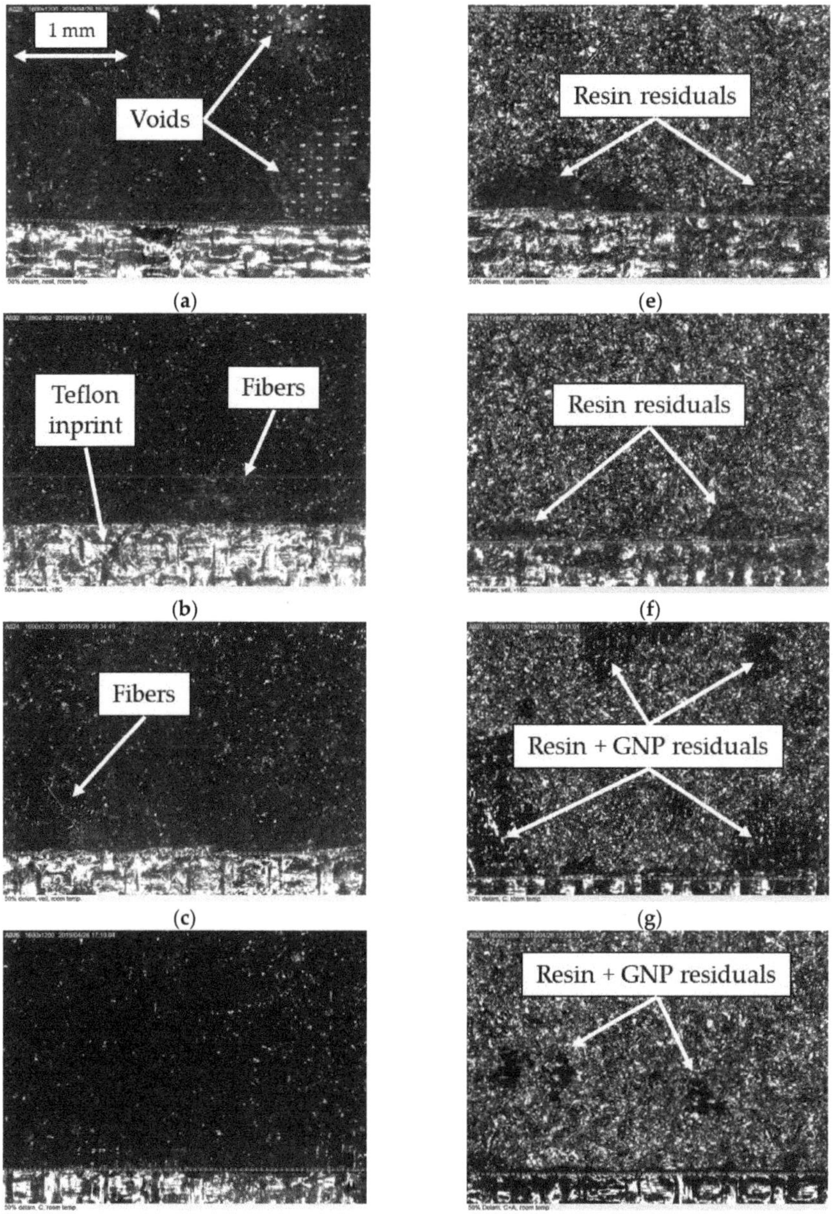

Figure 14. Magnified views of the adherends' fracture surfaces for specimens of case study II. Images (**a**) to (**d**) show the adhesive interfacial surfaces, while images (**e**) to (**h**) show the magnesium interfacial surfaces. From left to right: specimens "N", "V", "C" and "CA".

Moreover, the relatively small regions of resin residuals visible on the magnesium interface surfaces (see Figure 14e) suggest that the delamination initiated and propagated mainly at the interface between the magnesium skins and the resin layer.

Please note the relatively darker colour regions of resin residuals that can be seen in two cases (i.e., Figure 14g,h). The two cases correspond to the specimens that were prepared by the SBC bonding method and containing GNPs (i.e., specimens C and CA). The darker colour is believed to represent regions with a higher concentration (agglomeration) of nanoparticles developed in specimens that contained GNP only in the coating and in both coating and adhesive, respectively.

In this case, the darker magnesium bonding surfaces reported for case study I are mitigated using the new bonding method and incorporation of the cold cure adhesive, which is believed to have improved the interface compatibility, thus increasing the interface strength of the specimens of this case study. However, even with the new surface preparation method, the failure mode remains as an interfacial type. In comparison, more consistent and relatively substantive improvements could be gained by the inclusion of the more cost-effective fiberglass veil in the interface.

3.3. Case Study III

The last case study aims to investigate the effect of sub-freezing temperature on the performances of the 3D-FML hosting a delamination and whether the interfacial delamination resistance could be enhanced by the inclusion of GNPs. For this, the specimens of this case study were tested at a quasi-static loading rate of 0.5 mm/min. The imposed displacement of 1.4 mm facilitated the desired state of delamination propagation without causing the complete failure of the specimens (similar to what was done in the second case study). As stated earlier, the sub-freezing environment of this case study was generated by using liquid nitrogen, hence, the specimens of this case study are referred to as the LN2 specimens, and the specimens tested at room temperature are referred to as RT specimens.

The qualitative response of the specimens was identical to the behaviour described for the specimens of case study II as was illustrated in Figure 12; therefore, for the sake of space, the images are not presented. However, the quantitative results of the static buckling tests conducted both at room and sub-freezing temperatures are reported in Figures 15 and 16. The results illustrated in Figure 15 reveal that the LN2 specimens show vary similar stiffness compared with the RT specimens. However, the load-bearing capacity seems to be slightly higher for the LN2 specimens. Moreover, no clear distinction between the responses of neat and GNP-reinforced specimens can be seen, except for the case of LN2-CA specimens, which show slightly higher stiffness compared to the other LN2 specimens. Moreover, similar to the performance of specimens of case study II, the specimens hosting the fiberglass veil exhibited the best performances amongst the tested specimens in terms of buckling capacity at both room and sub-freezing temperatures.

To facilitate an easier comparison, the normalized buckling load capacities are reported in Figure 16a. The values are normalized with respect to the average value corresponding to the neat specimens tested at room-temperature (RT-N). The buckling load was taken as the load at which the linear slope of the load-displacement curve changes to a non-linear one (see point B on the graphs of Figure 15). The results also reveal that the inclusion of nanoparticles had a negligible effect on the buckling capacity of the specimens tested at both temperatures, reaching a maximum of approximately 5% for the RT-CA specimens. On the contrary, the more cost-effective inclusion of the veil within the interface increased the buckling capacity by 12% and 22%, respectively for specimens tested at RT and −50 °C, respectively.

In addition, the normalized delamination propagation response of the specimens are reported in Figure 16b (results normalized with respect to the RT-N case). The sub-freezing temperature caused the delamination to grow to a greater length compared to the response observed at RT. Specifically, the delamination length increased by 48%, 35%, 100%, and 78% for the specimens of categories neat, veil, C, and CA, respectively. Interestingly, while the presence of the interface veil reduced the growth of delamination by 28% when specimens were tested at RT, the veil's effect diminished significantly in specimens that were tested at the sub-freezing temperature; nonetheless, the veil still helped to suppress the delamination growth when compared to the growth observed in specimens that did not have the veil at their interfaces. It can also be seen that although the test results (i.e., load-axial shortening curves) are very consistent and have very low standard deviations, nevertheless, the standard deviations are relatively large when considering the delamination length results. This observation further validates our earlier statement that such large standard deviations are inherent to delamination growth being an unstable phenomenon in brittle materials. Also, similar to the results seen in the other case studies, the use of the veil resulted in the highest overall buckling capacity and the highest delamination mitigation, with the proviso that its effectiveness becomes adversely impacted by the sub-freezing temperature.

Finally, please note that case study III's bonding surface micrographs are omitted because they were very similar to those shown in Figure 14, thus not further information would be provided.

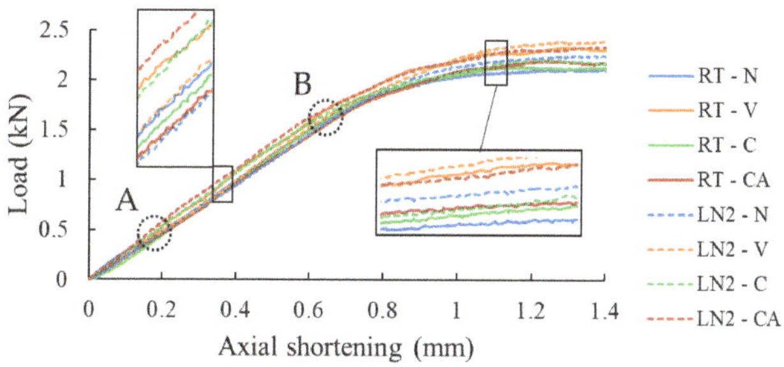

Figure 15. Axial load vs. axial shortening curves. Point A corresponds to the onset of buckling of the delaminated skin, while point B corresponds to the onset of the global buckling of the specimen.

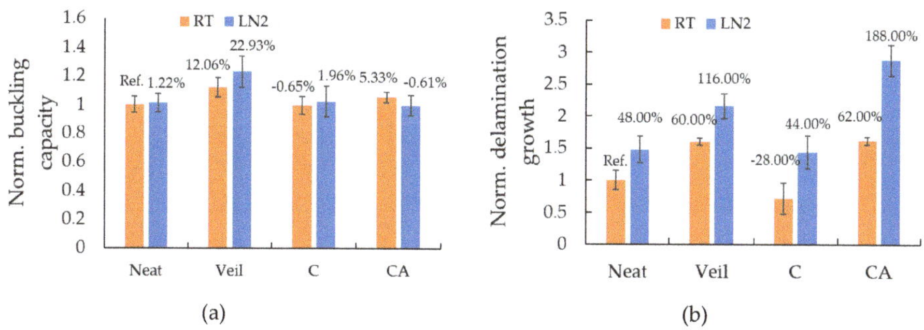

Figure 16. (**a**) normalized buckling capacity and (**b**) normalized delamination growth length for specimens tested at room and −50 °C.

4. Summary and Conclusions

A systematic investigation was conducted to examine the effect of graphene nanoplatelets (GNPs) used to reinforce a structural epoxy resin. The resin was used to mate the magnesium skins and composite core of a recently developed 3D fiber-metal laminate (3D-FML). The response of the resin and interface strength in the 3D-FML specimens were evaluated by subjecting the specimens to compressive loading at quasi-static and impact loading rates. Therefore, the impact buckling strength, delamination buckling strength, and delamination propagation were used as the evaluation criteria in this study. Two different techniques were used to join the skins to the FRP core. In the first method, the skins were directly bonded to the core using a hot-cure structural resin, with the mating skins' surfaces prepared by the conventional abrasive (sandblasting) method. In the second method, a cold-cure less expensive structural resin was used, and a newly developed resin coating method was employed for preparing the skins' mating surface. The specimens prepared using the first technique (i.e., case study I specimens), were axially impacted at four energies (1.5 J. 3 J, 4.5 J and 7 J). Two different case studies were organized to examine the effect of initial delamination present in such 3D-FMLs by considering intact specimens and specimens with a delamination length of 30%, 50% and 70% (percentiles refer to the ratio of delamination length to specimen's gage length). Moreover, GNP contents of 0.5 wt%, 1 wt% and 2 wt% were used to reinforce the resin in this study. The results from the first case study can be summarized as follows:

- The presence of initial delamination greatly affected the load-bearing capacity of the specimens, but its length had a negligible effect.
- For the intact specimens (i.e., with no initial delamination), the incorporation of GNPs showed its maximum enhancing effect when the specimens were subjected to the highest impact energy (7 J). The observed enhancements were 12.5%, 10.9%, and 10.7% corresponding to GNP contents of 0.5 wt%, 1 wt%, and 2 wt%, respectively. Ironically, a degradation of the strength was noted in specimens that were subjected to 4.5 J impact energy.
- Among the specimens that hosted a delamination, the specimens that were reinforced with 0.5 wt% of GNP content exhibited the most gain in strength under three out of the four impact energies tried. The exceptions were the specimens that were subjected to 4.5 J impact energy, for which 2 wt% GNP content produced the best results.
- Microscopic examination revealed the existence of some voids at the bonding interface of the 3D-FMLs.

To further explore the effect of GNP inclusion on the performance of the magnesium/FRP interface (and overall 3D-FML), additional case studies were considered. In the second case study, the specimens had the optimum GNP content of 0.5 wt%, with a fixed initial delamination of 50%, all tested under 2.85 J impact energy. The outcome of this case study is summarized as follows:

- The delamination propagated in an unstable manner.
- A higher GNP content led to a higher delamination length, with a 100% increase in delamination growth observed in the CA specimens.
- The use of a fiberglass veil interleaved between the magnesium and the FRP core mitigated the delamination extension by an average of 46% and increased the load-bearing capacity by 6%.
- The GNPs inclusion produced either no effect on the load capacity of most specimens or led to even negative effect in some (a reduction of 8% was observed in the CA specimens).
- The void content in the bonding region was drastically reduced when the SBC method was employed and voids were completely nullified when the veil or GNPs were incorporated within the interface; nonetheless, the delamination growth persisted owing to the lack of optimal chemical compatibility between magnesium and epoxy resin.

Finally, an investigation was carried out in a case study (III) examining the effect of sub-freezing temperature (−50 °C) on the delamination buckling and propagation of the 3D-FML and the effect of GNP inclusion. The specimens within this case study were tested under a quasi-static loading rate. The results are summarized as follows:

- The specimens' apparent stiffness changed marginally when exposed to the sub-freezing temperature.
- The buckling load capacity was positively affected by the sub-freezing temperature, especially when the veil was used.
- The sub-freezing environment caused an increase in delamination growth, especially in the GNP-reinforced specimens.

Overall, it can be concluded that some improvement in performances could be gained by incorporating GNPs in the interface of the 3D-FMLs; however, one could also expect degradation of the performance under certain circumstances. In comparison, incorporation of the fiberglass veil as demonstrated in this study would be a more effective and less costly means for enhancing the performance of 3D-FMLs under in-plane compressive loading. Not only is the cost of the veil lower than that of GNPs, but the labor cost associated with its incorporation would be much less than that required for processing GNPs into the resin.

In closing, the lack of the expected gain in performance as a result of reinforcing the resin with GNP is believed to be due to the lack of chemical compatibility between the resin and magnesium. The incompatibility does not allow the GNPs to demonstrate their full potential in enhancing the strength of the interface resin. This is mainly because the failure along the interface is in the interfacial mode (failure or resin/magnesium interface), as opposed to being of a cohesive type. Therefore, it is strongly believed that future works should focus on improving the chemical compatibility between the resin and magnesium. Based on the results of this study and those reported in the literature, it is strongly believed that once the interface compatibility issue is resolved, the incorporation of nanoparticles will positively and significantly influence the interface strength and hence the overall performance of 3D-FMLs when subject to in-plane loadings.

Author Contributions: D.D.C fabricated and tested the specimens, as well as conducting the experiments and the analysis of the data. D.D.C. also wrote the draft of the manuscript. F.T. conceptualized the research, secured funding and supervised and guided the overall research and edited the manuscript.

Funding: This research was supported by the Natural Sciences and Engineering Research Council of Canada (NSERC) and the MITACS Globalink fellowship program.

Acknowledgments: The grants received from the above agencies facilitated the study; the authors are indebted to these agencies. The authors are also grateful to Andrew Corkum (Dalhousie University) for kindly providing the Chronos high-speed camera, and Stephen Corbin (Dalhousie University) for kindly providing the liquid nitrogen tank and connection jig.

Conflicts of Interest: The authors declare no conflict of interest.

Abbreviations and Acronyms

3D-FGF	3D fiberglass fabric
3D-FML	3D fiber-metal laminate
C	nanoparticles included in the coating
CA	nanoparticles included in the coating and the adhesive
CNT	carbon nano-tubes
FML	fiber-metal laminate
FRP	fiber-reinforced polymer
GNP	graphene nanoplatelets
LN2	liquid nitrogen
N	specimens with neat resin
ND	no initial delamination
NP	nanoparticles
RMS	root-mean square
RT	room temperature
V	fiberglass veil
wt%	weight percentage
x%	percentage of initial delamination
Note	"s" following above acronyms make them plural

References

1. Kim, J.T.; Hong, S.H.; Park, H.J.; Kim, Y.S.; Suh, J.Y.; Lee, J.K.; Park, J.M.; Maity, T.; Eckert, J.; Kim, K.B. Deformation mechanisms to ameliorate the mechanical properties of novel TRIP/TWIP Co-Cr-Mo-(Cu) ultrafine eutectic alloys. *Sci. Rep.* **2017**, *7*, 39959. [CrossRef] [PubMed]

2. Kim, J.T.; Hong, S.H.; Park, H.J.; Park, G.H.; Suh, J.Y.; Park, J.M.; Kim, K.B. Influence of microstructural evolution on mechanical behavior of Fe-Nb-B ultrafine composites with a correlation to elastic modulus and hardness. *J. Alloys Compd.* **2015**, *647*, 886–891. [CrossRef]
3. Asaee, Z.; Shadlou, S.; Taheri, F. Low-velocity impact response of fiberglass/magnesium FMLs with a new 3D fiberglass fabric. *Compos. Struct.* **2015**, *122*, 155–165. [CrossRef]
4. Asaee, Z.; Taheri, F. Experimental and numerical investigation into the influence of stacking sequence on the low-velocity impact response of new 3D FMLs. *Compos. Struct.* **2016**, *140*, 136–146. [CrossRef]
5. De Cicco, D.; Taheri, F. Delamination Buckling Response of 3D Fiber-Metal Laminates Subjected to Different Loading Rates. In Proceedings of the American Society for Composites: Thirty-First Technical Conference, Williamsburg, VA, USA, 19–23 September 2016; DEStech Publications, Inc.: Lancaster, PA, USA, 2016; p. 12.
6. De Cicco, D.; Taheri, F. Robust numerical approaches for simulating the buckling response of 3D fiber-metal laminates under axial impact—Validation with experimental results. *J. Sandw. Struct. Mater.* **2018**. [CrossRef]
7. De Cicco, D.; Taheri, F. Understanding the parameters that influence buckling of 3D fiber-metal laminates. In Proceedings of the 10th Canadian-International Conference on Composites (CANCOM2017), Ottawa, ON, Canada, 17–20 July 2017.
8. Shahid, M.; Hashim, S.A. Effect of surface roughness on the strength of cleavage joints. *Int. J. Adhes. Adhes.* **2002**, *22*, 235–244. [CrossRef]
9. Baldan, A. Adhesively-bonded joints and repairs in metallic alloys, polymers and composite materials: Adhesives, adhesion theories and surface pretreatment. *J. Mater. Sci.* **2004**, *39*, 1–49. [CrossRef]
10. Harris, A.F.; Beevers, A. The ffects of grit-blasting on surface properties for adhesion. *Int. J. Adhes. Adhes.* **1999**, *19*, 445–452. [CrossRef]
11. Williams, T.S. *Surface Modification by Atmospheric Pressure Plasma for Improved Bonding*; University of California: Los Angeles, CA, USA, 2013.
12. Gonzalez, E.; Hicks, R.F. Surface Analysis of Polymers Treated by Remote Atmospheric Pressure Plasma. *Langmuir* **2009**, *26*, 3710–3719. [CrossRef]
13. Livadifiis, G.J. A chemical etching system for creating micromechanical retention in resin-bonded retainers. *J. Prosthet. Dent.* **1986**, *56*, 181–188. [CrossRef]
14. Lefebvre, D.R.; Ahn, B.K.; Dillard, D.A.; Dillard, J.G. The effect of surface treatments on interfacial fatigue crack initiation in aluminum/epoxy bonds. *Int. J. Fract.* **2002**, *114*, 191–202. [CrossRef]
15. Wang, B.; Hu, X.; Lu, P. Improvement of adhesive bonding of grit-blasted steel substrates by using diluted resin as a primer. *Int. J. Adhes. Adhes.* **2017**, *73*, 92–99. [CrossRef]
16. Javni, I.; Zhang, W.; Karajkov, V.; Petrovic, Z.S.; Divjakovic, V. Effect of Nano- and Micro-Silica Fillers on Polyurethane Foam Properties. *J. Cell. Plast.* **2002**, *38*, 229–239. [CrossRef]
17. Rong, M.Z.; Zhang, M.Q.; Zheng, Y.X.; Zeng, H.M.; Friedrich, K. Improvement of tensile properties of nano-SiO2/PP composites in relation to percolation mechanism. *Polymer (Guildf.)* **2001**, *42*, 3301–3304. [CrossRef]
18. Jalili, M.M.; Moradian, S.; Dastmalchian, H.; Karbasi, A. Investigating the variations in properties of 2-pack polyurethane clear coat through separate incorporation of hydrophilic and hydrophobic nano-silica. *Prog. Org. Coat.* **2007**, *59*, 81–87. [CrossRef]
19. Ji, G.; Li, G. Effects of nanoclay morphology on the mechanical, thermal, and fire-retardant properties of vinyl ester based nanocomposite. *Mater. Sci. Eng. A* **2008**, *498*, 327–334. [CrossRef]
20. Park, J.H.; Jana, S.C. The relationship between nano-and micro-structures and mechanical properties in PMMA–epoxy–nanoclay composites. *Polymer (Guildf.)* **2003**, *44*, 2091–2100. [CrossRef]
21. Mohan, T.P.; Kanny, K. Water barrier properties of nanoclay filled sisal fibre reinforced epoxy composites. *Compos. Part A Appl. Sci. Manuf.* **2011**, *42*, 385–393. [CrossRef]
22. Dzenis, Y.A.; Reneker, D.H. Delamination Resistant Composites Prepared by Small Diameter Fiber Reinforcement at Ply Interfaces. US Patent Number 6,265,333, 24 July 2001.
23. van der Heijden, S.; De Bruycker, K.; Simal, R.; Du Prez, F.; De Clerck, K. Use of Triazolinedione Click Chemistry for Tuning the Mechanical Properties of Electrospun SBS-Fibers. *Macromolecules* **2015**, *48*, 6474–6481. [CrossRef]
24. Kucheryavy, P.; He, J.; John, V.T.; Maharjan, P.; Spinu, L.; Goloverda, G.Z.; Kolesnichenko, V.L. Superparamagnetic iron oxide nanoparticles with variable size and an iron oxidation state as prospective imaging agents. *Langmuir* **2013**, *29*, 710–716. [CrossRef]

25. Sardana, S.K.; Chava, V.S.N.; Thouti, E.; Chander, N.; Kumar, S.; Reddy, S.R.; Komarala, V.K. Influence of surface plasmon resonances of silver nanoparticles on optical and electrical properties of textured silicon solar cell. *Appl. Phys. Lett.* **2014**, *104*, 073903. [CrossRef]
26. Chou, C.-W.; Hsu, S.-H.; Chang, H.; Tseng, S.-M.; Lin, H.-R. Enhanced thermal and mechanical properties and biostability of polyurethane containing silver nanoparticles. *Polym. Degrad. Stab.* **2006**, *91*, 1017–1024. [CrossRef]
27. Domun, N.; Hadavinia, H.; Zhang, T.; Sainsbury, T.; Liaghat, G.H.; Vahid, S. Improving the fracture toughness and the strength of epoxy using nanomaterials—A review of the current status. *Nanoscale* **2015**, *7*, 10294–10329. [CrossRef] [PubMed]
28. Borowski, E.; Soliman, E.; Kandil, U.F.; Taha, M.R. Interlaminar Fracture Toughness of CFRP Laminates Incorporating Multi-Walled Carbon Nanotubes. *Polymers (Basel)* **2015**, *7*, 1020–1045. [CrossRef]
29. Chandrasekaran, S.; Seidel, C.; Schulte, K. Preparation and characterization of graphite nano-platelet (GNP)/ epoxy nano-composite: Mechanical, electrical and thermal properties. *Eur. Polym. J.* **2013**, *49*, 3878–3888. [CrossRef]
30. Eskizeybek, V.; Avci, A.; Gülce, A. The Mode I interlaminar fracture toughness of chemically carbon nanotube grafted glass fabric/epoxy multi-scale composite structures. *Compos. Part A Appl. Sci. Manuf.* **2014**, *63*, 94–102. [CrossRef]
31. Wang, P.; Liu, W.; Zhang, X.; Lu, X.; Yang, J. Enhanced fracture toughness of carbon fabric/epoxy laminates with pristine and functionalized stacked-cup carbon nanofibers. *Eng. Fract. Mech.* **2015**, *148*, 73–81. [CrossRef]
32. Shokrieh, M.M.; Ghoreishi, S.M.; Esmkhani, M.; Zhao, Z. Effects of graphene nanoplatelets and graphene nanosheets on fracture toughness of epoxy nanocomposites. *Fatigue Fract. Eng. Mater. Struct.* **2014**, *37*, 1116–1123. [CrossRef]
33. Alishahi, E.; Shadlou, S.; Doagou-R, S.; Ayatollahi, M.R. Effects of carbon nanoreinforcements of different shapes on the mechanical properties of epoxy-based nanocomposites. *Macromol. Mater. Eng.* **2013**, *298*, 670–678. [CrossRef]
34. Ayatollahi, M.R.; Shadlou, S.; Shokrieh, M.M.; Chitsazzadeh, M. Effect of multi-walled carbon nanotube aspect ratio on mechanical and electrical properties of epoxy-based nanocomposites. *Polym. Test.* **2011**, *30*, 548–556. [CrossRef]
35. Shen, M.-Y.; Chang, T.-Y.; Hsieh, T.-H.; Li, Y.-L.; Chiang, C.-L.; Yang, H.; Yip, M.-C. Mechanical Properties and Tensile Fatigue of Graphene Nanoplatelets Reinforced Polymer Nanocomposites. *J. Nanomater.* **2013**, *2013*, 1–9. [CrossRef]
36. Yavari, F.; Rafiee, M.A.; Rafiee, J.; Yu, Z.-Z.; Koratkar, N. Dramatic increase in fatigue life in hierarchical graphene composites. *ACS Appl. Mater. Interfaces* **2010**, *2*, 2738–2743. [CrossRef] [PubMed]
37. Khan, S.U.; Li, C.Y.; Siddiqui, N.A.; Kim, J.-K. Vibration damping characteristics of carbon fiber-reinforced composites containing multi-walled carbon nanotubes. *Compos. Sci. Technol.* **2011**, *71*, 1486–1494. [CrossRef]
38. DeValve, C.; Pitchumani, R. Experimental investigation of the damping enhancement in fiber-reinforced composites with carbon nanotubes. *Carbon* **2013**, *63*, 71–83. [CrossRef]
39. DeValve, C.; Pitchumani, R. Analysis of vibration damping in a rotating composite beam with embedded carbon nanotubes. *Compos. Struct.* **2014**, *110*, 289–296. [CrossRef]
40. Liu, A.; Wang, K.W.; Bakis, C.E. Effect of functionalization of single-wall carbon nanotubes (SWNTs) on the damping characteristics of SWNT-based epoxy composites via multiscale analysis. *Compos. Part A Appl. Sci. Manuf.* **2011**, *42*, 1748–1755. [CrossRef]
41. Soltannia, B.; Haji Gholami, I.; Masajedian, S.; Mertiny, P.; Sameoto, D.; Taheri, F. Parametric Study of Strain Rate Effects on Nanoparticle-Reinforced Polymer Composites. *J. Nanomater.* **2016**, *2016*, 1–9. [CrossRef]
42. Soltannia, B.; Taheri, F. Static, Quasi-Static and High Loading Rate Effects on Graphene Nano-Reinforced Adhesively Bonded Single-Lap Joints. *Int. J. Compos. Mater.* **2013**, *2013*, 181–190. [CrossRef]
43. Haro, E.E.; Odeshi, A.G.; Szpunar, J.A. The energy absorption behavior of hybrid composite laminates containing nano-fillers under ballistic impact. *Int. J. Impact Eng.* **2016**, *96*, 11–22. [CrossRef]
44. Ávila, A.F.; Neto, A.S.; Nascimento, H., Jr. Hybrid nanocomposites for mid-range ballistic protection. *Int. J. Impact Eng.* **2011**, *38*, 669–675. [CrossRef]

45. Haq, M.; Umer, R.; Khomenko, A.; Loos, A.C.; Drzal, L.T. Manufacturing and impact behavior of sandwich composites with embedded graphene platelets. In Proceedings of the 19th International Conference on Composite Materials ICCM19, Montreal, QC, Canada, 28 July–2 August 2014. [CrossRef]
46. Rafiee, M.A.; Rafiee, J.; Srivastava, I.; Wang, Z.; Song, H.; Yu, Z.Z.; Koratkar, N. Fracture and fatigue in graphene nanocomposites. *Small* **2010**, *6*, 179–183. [CrossRef] [PubMed]
47. Chandrasekaran, S.; Sato, N.; Tölle, F.; Mülhaupt, R.; Fiedler, B.; Schulte, K. Fracture toughness and failure mechanism of graphene based epoxy composites. *Compos. Sci. Technol.* **2014**, *97*, 90–99. [CrossRef]
48. Rafiee, M.A.; Rafiee, J.; Wang, Z.; Song, H.; Yu, Z.; Koratkar, N. Enhanced Mechanical Properties of Nanocomposites at Low Graphene Content. *ACS Nano* **2009**, *3*, 3884–3890. [CrossRef] [PubMed]
49. Ahmadi-Moghadam, B.; Taheri, F. Influence of graphene nanoplatelets on modes I, II and III interlaminar fracture toughness of fiber-reinforced polymer composites. *Eng. Fract. Mech.* **2015**, *143*, 97–107. [CrossRef]
50. Ahmadi-Moghadam, B.; Sharafimasooleh, M.; Shadlou, S.; Taheri, F. Effect of functionalization of graphene nanoplatelets on the mechanical response of graphene/epoxy composites. *Mater. Des.* **2015**, *66*, 142–149. [CrossRef]
51. Wichmann, M.H.G.; Sumfleth, J.; Gojny, F.H.; Quaresimin, M.; Fiedler, B.; Schulte, K. Glass-fibre-reinforced composites with enhanced mechanical and electrical properties—Benefits and limitations of a nanoparticle modified matrix. *Eng. Fract. Mech.* **2006**, *73*, 2346–2359. [CrossRef]
52. Siegfried, M.; Tola, C.; Claes, M.; Lomov, S.V.; Verpoest, I.; Gorbatikh, L. Impact and residual after impact properties of carbon fiber/epoxy composites modified with carbon nanotubes. *Compos. Struct.* **2014**, *111*, 488–496. [CrossRef]
53. Bortz, D.R.; Heras, E.G.; Martin-Gullon, I. Impressive Fatigue Life and Fracture Toughness Improvements in Graphene Oxide/Epoxy Composites. *Macromolecules* **2012**, *45*, 238–245. [CrossRef]
54. Remmers, J.J.C.; de Borst, R. Delamination buckling of fibre-metal laminates. *Compos. Sci. Technol.* **2001**, *61*, 2207–2213. [CrossRef]
55. Yin, S.; Yu, T.; Bui, T.Q.; Liu, P.; Hirose, S. Buckling and vibration extended isogeometric analysis of imperfect graded Reissner-Mindlin plates with internal defects using NURBS and level sets. *Comput. Struct.* **2016**, *177*, 23–38. [CrossRef]
56. Gong, W.; Chen, J.; Patterson, E.A. Buckling and delamination growth behaviour of delaminated composite panels subject to four-point bending. *Compos. Struct.* **2016**, *138*, 122–133. [CrossRef]
57. Gu, H.; Chattopadhyay, A. An experimental investigation of delamination buckling and post buckling of composite laminates. *Compos. Sci. Technol.* **1999**, *59*, 903–910. [CrossRef]
58. Wang, S.; Harvey, C.M.; Wang, B.; Watson, A. Post-local buckling-driven delamination in bilayer composite beams. *Compos. Struct.* **2015**, *133*, 1058–1066. [CrossRef]
59. Esfahani, M.M.N.; Ghasemnejad, H.; Barrington, P.E. Experimental and numerical buckling analysis of delaminated hybrid composite beam structures. *Appl. Mech. Mater.* **2010**, *24–25*, 393–400. [CrossRef]
60. Kim, H.-J.; Hong, C.-S. Buckling and Postbuckling Behavior of Composite Laminates with a Delamination. *Compos. Sci. Technol.* **1997**, *57*, 557–564. [CrossRef]
61. Asaee, Z.; Mohamed, M.; Soumik, S.; Taheri, F. Experimental and numerical characterization of delamination buckling behavior of a new class of GNP-reinforced 3D fiber-metal laminates. *Thin Walled Struct.* **2017**, *112*, 208–216. [CrossRef]
62. Taraghi, I.; Fereidoon, A.; Taheri-Behrooz, F. Low-velocity impact response of woven Kevlar/epoxy laminated composites reinforced with multi-walled carbon nanotubes at ambient and low temperatures. *Mater. Des.* **2014**, *53*, 152–158. [CrossRef]
63. Shen, X.-J.; Meng, L.-X.; Yan, Z.-Y.; Sun, C.-J.; Ji, Y.-H.; Xiao, H.-M.; Fu, S.-Y. Improved cryogenic interlaminar shear strength of glass fabric/epoxy composites by graphene oxide. *Compos. Part B Eng.* **2015**, *73*, 126–131. [CrossRef]
64. Müller, B.; Sinke, J.; Anisimov, A.G.; Groves, R.M. Thermal Strains in Heated Fiber Metal Laminates. In Proceedings of the Conference on Emerging Technologies in Non-Destructive Testing (ETNDT6), Brussels, Belgium, 27–29 May 2015.
65. Li, H.; Hu, Y.; Liu, C.; Zheng, X.; Liu, H.; Tao, J. The effect of thermal fatigue on the mechanical properties of the novel fiber metal laminates based on aluminum-lithium alloy. *Compos. Part A Appl. Sci. Manuf.* **2016**, *84*, 36–42. [CrossRef]

66. Khalili, S.M.R.; Sharafi, M.; Eslami-Farsani, R.; Saeedi, A. Effect of thermal cycling on tensile properties of degraded FML to metal hybrid joints exposed to sea water. *Int. J. Adhes. Adhes.* **2017**, *79*, 95–101. [CrossRef]
67. De Cicco, D.; Taheri, F. Enhancement of magnesium-composite bond-interface by a simple combined abrasion and coating method. *J. Magnes. Alloys* **2019**, *7*, 227–239. [CrossRef]

© 2019 by the authors. Licensee MDPI, Basel, Switzerland. This article is an open access article distributed under the terms and conditions of the Creative Commons Attribution (CC BY) license (http://creativecommons.org/licenses/by/4.0/).

nanomaterials

MDPI

Article

An Ab Initio Study of Pressure-Induced Reversal of Elastically Stiff and Soft Directions in YN and ScN and Its Effect in Nanocomposites Containing These Nitrides

Martin Friák [1,*], Pavel Kroupa [1,2], David Holec [3] and Mojmír Šob [4,1,5]

1 Institute of Physics of Materials, Academy of Sciences of the Czech Republic, Žižkova 22, CZ-616 62 Brno, Czech Republic; kroupavel@gmail.com (P.K.); mojmir@ipm.cz (M.Š.)
2 Department of Physics, Imperial College London, Prince Consort Road, London SW7 2BP, UK
3 Department of Materials Science, Montanuniversität Leoben, Franz-Josef-Strasse 18, A-8700 Leoben, Austria; david.holec@unileoben.ac.at
4 Department of Chemistry, Faculty of Science, Masaryk University, Kotlářská 2, CZ-611 37 Brno, Czech Republic
5 Central European Institute of Technology, CEITEC MU, Masaryk University, Kamenice 5, CZ-625 00 Brno, Czech Republic
* Correspondence: friak@ipm.cz; Tel.: +420-532-290-400

Received: 22 November 2018; Accepted: 11 December 2018; Published: 14 December 2018

Abstract: Using quantum-mechanical calculations of second- and third-order elastic constants for YN and ScN with the rock-salt (B1) structure, we predict that these materials change the fundamental type of their elastic anisotropy by rather moderate hydrostatic pressures of a few GPa. In particular, YN with its zero-pressure elastic anisotropy characterized by the Zener anisotropy ratio $A_Z = 2C_{44}/(C_{11} - C_{12}) = 1.046$ becomes elastically isotropic at the hydrostatic pressure of 1.2 GPa. The lowest values of the Young's modulus (so-called soft directions) change from $\langle 100 \rangle$ (in the zero-pressure state) to the $\langle 111 \rangle$ directions (for pressures above 1.2 GPa). It means that the crystallographic orientations of stiffest (also called hard) elastic response and those of the softest one are reversed when comparing the zero-pressure state with that for pressures above the critical level. Qualitatively, the same type of reversal is predicted for ScN with the zero-pressure value of the Zener anisotropy factor $A_Z = 1.117$ and the critical pressure of about 6.5 GPa. Our predictions are based on both second-order and third-order elastic constants determined for the zero-pressure state but the anisotropy change is then verified by explicit calculations of the second-order elastic constants for compressed states. Both materials are semiconductors in the whole range of studied pressures. Our phonon calculations further reveal that the change in the type of the elastic anisotropy has only a minor impact on the vibrational properties. Our simulations of biaxially strained states of YN demonstrate that a similar change in the elastic anisotropy can be achieved also under stress conditions appearing, for example, in coherently co-existing nanocomposites such as superlattices. Finally, after selecting ScN and PdN (both in B1 rock-salt structure) as a pair of suitable candidate materials for such a superlattice (due to the similarity of their lattice parameters), our calculations of such a coherent nanocomposite results again in a reversed elastic anisotropy (compared with the zero-pressure state of ScN).

Keywords: YN; ScN; pressure; elasticity; ab initio; stability; nanocomposites

1. Introduction

Anisotropic (tensorial) elastic characteristics belong to the most fundamental properties of crystals (see [1,2]) and reflect the nature of inter-atomic bonds. Elastic constants are decisive for numerous phenomena well beyond simple mechanical response of crystal lattices to uniaxial, biaxial or triaxial loading. In particular, strong long-range elastic interactions among point defects, such as substitutional or interstitial solute atoms, are inter-linked with their low solubilities (see [3]). Further, elastic interactions among extended defects, such as edge or screw dislocations [4], grain boundaries or stacking faults, are crucial for phenomena mediating plasticity in crystalline materials. As another example, when changing the temperature in composites, it is the varying lattice-parameter mismatch and elastic stiffnesses of coexisting phases which play an important role in stresses occurring at internal interfaces. Regarding the mechanical stability and the very existence of materials phases, relations among elastic constants are vital for the mechanical stability as violations of so-called Born-Huang stability criteria [5] often lead to phase transformations. As it would be indeed very difficult to provide a complete list of phenomena intertwined with elastic properties, we further mention only their role in sound propagation [6–10], heat transfer and partly also in thermal lattice vibrations [11], which significantly contribute to the thermodynamic stability (phonon entropy contribution is a part of the free energy [12]).

When considering elastic anisotropy of materials, its most important characteristics are the magnitude of the anisotropy (as a measure of how different the elastic response is from the isotropic one) and then special directions for which the studied crystalline system exhibits the softest and stiffest elastic response. Focusing on crystals with a cubic symmetry, which are described by three independent elastic constants (stiffnesses) C_{11}, C_{12} and C_{44}, the magnitude of the elastic anisotropy is often expressed by the so-called Zener's anisotropy ratio $A_Z = 2C_{44}/(C_{11} - C_{12})$. If this ratio exceeds one, the elastically stiffest response of the studied cubic crystal to uniaxial loading is found along the $\langle 111 \rangle$ family of directions while the softest response occurs along the $\langle 001 \rangle$ directions. When the Zener's ratio is lower than one, the elastic anisotropy is opposite and the stiffest elastic response, i.e., the highest value of the Young's modulus, is found along the $\langle 001 \rangle$ directions. The border case of $A_Z = 1$ describes an elastically isotropic material (the directional dependence of Young's modulus would be a sphere).

Together with the above discussed second-order elastic constants, there are also elastic constants related to higher-order elasticity [13–16]. In particular, there are six independent third-order elastic constants in the case of cubic-symmetry systems: $C_{111} = C_{222} = C_{333}$, $C_{144} = C_{255} = C_{366}$, $C_{112} = C_{223} = C_{133} = C_{113} = C_{122} = C_{233}$, $C_{155} = C_{244} = C_{344} = C_{166} = C_{266} = C_{355}$, C_{123}, C_{456} and all other are zero (provided that the mutual orientation of the lattice and the coordination system is matching). Importantly, third-order elastic constants describe the changes of the second-order elastic constants due to the application of external stress or strain, including a hydrostatic pressure p.

In our study we show that the fundamental elastic anisotropy type, i.e., whether the elastically softest response is either along $\langle 001 \rangle$ or $\langle 111 \rangle$ directions and the corresponding Zener anisotropy ratio either higher or lower than 1, respectively, can be changed by application of moderate hydrostatic pressures. Importantly, the predicted reversal means that the mutual ratio of longitudinal sound velocities (which is faster or slower) in the [100] direction on one hand and in the [110] and [111] directions on the other, reverse as well [17]. Such a change has been reported mostly as a consequence of compositional changes so far, e.g., in Ref. [18], while we found that it caused by the hydrostatic pressure.

We predict the reversal for YN and ScN as two technologically important materials which have been intensively studied. Regarding YN, its electronic structure, vibrational spectrum, and thermal properties were computed using first-principles density functional theory (DFT) based simulations with a generalized gradient approximation (GGA) of the exchange correlation energy in [19]. The authors of that study also applied the Hubbard on-site correlation U term (GGA+U) and reported improvement in the accuracy of the calculation of the bandgap and selected features of the electronic structure of YN which are relevant to transport properties, such as transverse and longitudinal conduction band effective mass. The GGA+U calculations were also performed in the study of electronic, mechanical, and thermodynamic properties of YN in [20]. Other theoretical studies were focused on the stability of the rock-salt B1 structure of YN with respect to a pressure-induced transition into another phase, such as the caesium-chloride B2 one [21,22]. Full-potential calculations were also performed in the theoretical study by Stampfl and co-workers [23] who showed that local density approximation (LDA) predicts YN to be semimetal and the bandgap is open only when a screened-exchange calculations are performed. As far as ScN is concerned, it was a part of an extensive study of properties of 3d transition metal nitrides considering their cubic zinc-blende, rock-salt and caesium-chloride polymorphs [24]. It was also one of the compounds in the study of the Sc-based ternary nitrides [25] which was focused on single-crystal elastic constants, mechanical stability, the site-projected density of states, Fermi surfaces, charge densities and chemical bonding.

2. Methods

Our quantum-mechanical calculations were performed within the framework of density functional theory [26,27] using the Vienna Ab initio Simulation Package (VASP) [28,29] and projector augmented wave (PAW) pseudopotentials [30,31] (electron configuration in the VASP-notation Y-sv: 4s4p5s4d, Sc-sv: 3p4s3d, N: s2p3). The computational setting was the same as in Ref. [32]. The exchange and correlation energy was treated in the generalized gradient approximation (GGA) as parametrized by Perdew and Wang [33]. We used a plane-wave energy cut-off of 800 eV, a $7 \times 7 \times 7$ Monkhorst-Pack k-point mesh and 8-atom cube-shaped computational supercells (see a schematic visualization of this structure in Figure 1b). Second- and third-order elastic constants at zero pressure were computed as described in our paper [32] which also contains very detailed convergence tests. The second-order elastic constants under pressure (also in the case of tetragonal-symmetry states) were determined using the stress-strain method [34]. In this case, Born stability conditions in their original version are also valid for non-zero pressures and the external pressure does not enter here explicitly. In order to obtain highly accurate densities of states, $14 \times 14 \times 14$ k-point meshes were used in the case of the above discussed 8-atom cells. These calculations were initially performed with employing the Fermi smearing (VASP-parameter ISMEAR = -1) with the smearing parameter σ = 0.02 eV. After reaching a self-consistent solution for a given geometry (for each studied lattice parameter), a non-selfconsistent run (VASP-parameter ICHARG = 11) was subsequently performed utilizing the tetrahedron method (VASP-parameter ISMEAR = -5) to compute the density of states (following the VASP manual). For phonon calculations we have used 64-atom $2 \times 2 \times 2$ multiple of the cube-shape 8-atom elementary cell (which is shown in Figure 1b). The corresponding k-point mesh was then $4 \times 4 \times 4$. Phonopy [35] software package was utilized.

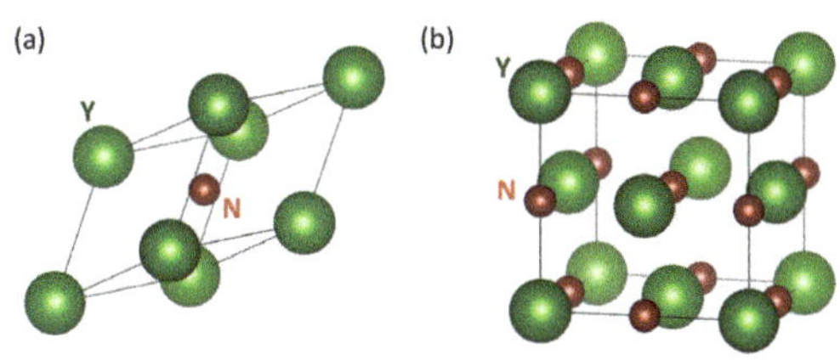

Figure 1. Schematic visualization of the 2-atom primitive (**a**) and 8-atom conventional cube-shape (**b**) unit cells of NaCl-structure of YN (some atoms are shown together with their periodic images).

3. Results

As far as the ground-state properties of B1-structure YN and ScN are concerned, the calculated equilibrium lattice parameters are in an excellent agreement with those previously obtained that employed different variants of the GGA exchange-correlation approximations as well as with experimental results. In particular, we find the lattice parameter of YN to be 4.916 Å when theoretical values 4.90–4.93 Å were reported in Reference [19], 4.619 Å in [20], 4.93 Å in [21,36], 4.85 Å in [23] and the experimental value is 4.88 Å [37]. Regarding ScN, our value 4.510 Å agrees with theoretical ones of 4.543 Å from [24], 4.516 Å reported in Reference [38], 4.50 Å in [23] and experimental 4.50 Å [39].

Regarding elastic properties, the first-order derivatives of the second-order elastic constants C_{ij} with respect to the hydrostatic pressure p are in the case of cubic systems equal to (according to Refs. [13–15,32,40]):

$$\frac{\delta C_{11}}{\delta p} = -\frac{2C_{11} + 2C_{12} + C_{111} + 2C_{112}}{C_{11} + 2C_{12}}, \quad (1)$$

$$\frac{\delta C_{12}}{\delta p} = -\frac{-C_{11} - C_{12} + C_{123} + 2C_{112}}{C_{11} + 2C_{12}}, \quad (2)$$

$$\frac{\delta C_{44}}{\delta p} = -\frac{C_{11} + 2C_{12} + C_{44} + C_{144} + 2C_{166}}{C_{11} + 2C_{12}}. \quad (3)$$

The calculated values of second-order $C_{ij}(p = 0 \text{ GPa})$ and third-order elastic constants $C_{ijk}(p = 0 \text{ GPa})$ determined for ground-state (i.e., zero hydrostatic pressure, $p = 0$ GPa) of YN and ScN are summarized in Table 1. The elastic constants were recently published in our previous work [32] but here we newly add also the changes $\delta C_{11}/\delta p$, $\delta C_{12}/\delta p$ and $\delta C_{44}/\delta p$ according to Equations (1)–(3). Regarding the calculated values of second-order elastic constants $C_{ij}(p = 0 \text{ GPa})$, Table 1 shows that they are in an excellent agreement with previously published theoretical results for both YN and ScN when we selected GGA calculations [24,25,38] (LDA predicts both materials to be metallic [23]).

Table 1. Calculated second-order elastic constants $C_{ij}(p = 0 \text{ GPa})$ (in comparison with selected literature values—when a GGA was used as in our case) and their pressure changes ($\delta C_{11}/\delta p$, $\delta C_{12}/\delta p$, and $\delta C_{44}/\delta p$) as approximatively evaluated for $\delta p = 1$ GPa from computed third-order elastic constants $C_{ijk}(p = 0 \text{ GPa})$. Theoretical values taken from Ref. [24] are related to GGA-PW91 approximation [33] similarly as in our case (marked by *), GGA-PW91 + U (marked by **), GGA-PBE [41] (marked by †) or GGA-PBE + U (marked by ††).

	C_{11}	C_{12}	C_{44}	$\delta C_{11}/\delta p$	$\delta C_{12}/\delta p$	$\delta C_{44}/\delta p$
YN	318	81	124	7.55	1.12	-0.70
	(321 [24] *)	(81 [24] *)	(124 [24] *)			
	(304 [24] **)	(76 [24] **)	(122 [24] **)			
	(317 [24] †)	(80 [24] †)	(123 [24] †)			
	(310 [24] ††)	(81 [24] ††)	(124 [24] ††)			
ScN	388	106	166	7.49	1.02	-0.51
	(399 [24])	(96 [24])	(158 [24])			
	(397 [25])	(131 [25])	(170 [25])			
	(354 [38])	(100 [38])	(170 [38])			
	C_{111}	C_{112}	C_{123}	C_{144}	C_{166}	C_{456}
YN	−4100	−160	180	180	−225	185
ScN	−5100	−190	260	200	−330	215

The second-order elastic constants of B1 structure YN for the zero-pressure case $C_{ij}(p = 0 \text{ GPa})$ can be neatly visualized in the form of directional dependence of Young's modulus $Y(p = 0 \text{ GPa})$ in Figure 2a. The Young' s modulus is a measure of the response of the studied system to an uniaxial loading along different directions and as such it reflects the elastic anisotropy. Young's modulus in Figure 2a is nearly spherical, i.e., the elastic elastic anisotropy of YN at the zero-pressure case is weak and the corresponding Zener ratio $A_Z = 1.046$. In order to graphically represent the third-order elastic constants $C_{ijk}(p = 0 \text{ GPa})$, we conveniently visualize pressure-induced changes in the Young's modulus for each direction. In particular, we show the difference between the Young's modulus in the pressurized case $Y(p = 1 \text{ GPa})$ and Young's modulus for the zero-pressure case $Y(p = 0 \text{ GPa})$, i.e.,

$Y(p = 1\text{ GPa}) - Y(p = 0\text{ GPa})$. The values of these changes (in GPa) are shown for each direction in Figure 2b while relative changes (when the difference $Y(p = 1\text{ GPa}) - Y(p = 0\text{ GPa})$ is for each direction divided by the $Y(p = 0\text{ GPa})$ along this direction) are visualized in Figure 2c.

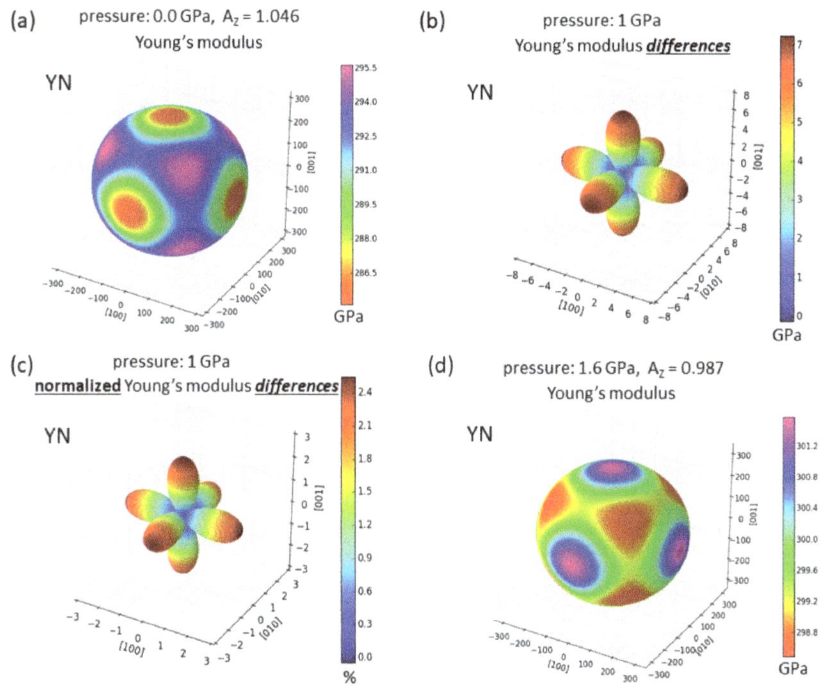

Figure 2. Computed changes in the elasticity of rock-salt structure YN visualized as directional dependencies of the Young's modulus. The zero-pressure case based on the second-order elastic constants computed by the stress-strain method is shown in part (**a**). The estimated changes in the Young's modulus due to 1 GPa of hydrostatic pressure are shown for different directions in absolute terms (in GPa) in part (**b**) and relatively (divided by the value for this direction in the zero-pressure case) in part (**c**). The visualized changes (in the second-order elasticity at the hydrostatic pressure of 1 GPa) are predicted using the second-order and third-order elastic constants computed for the zero-pressure state according to Equations (1)–(3). Finally, the directional dependence of the second-order elasticity computed at the 1.6 GPa is shown in part (**d**). Mind the change in the scale between the parts (**a**) and (**d**).

Figure 2b,c show that the Young's modulus is found to increase the most along the $\langle 001 \rangle$ directions. This is the direction along which the Young's modulus of YN in the zero-pressure case exhibits the softest elastic response (the lowest value, see Figure 2a). In contrast, the change for the the $\langle 111 \rangle$ directions is nearly zero (see Figure 2b,c). Figure 2b,c thus indicates that application of hydrostatic pressure can change the type of elastic anisotropy.

The Young's modulus of YN under pressure is predicted to have the stiffest (hard) elastic response along the $\langle 001 \rangle$ directions and not the softest one (as in the zero-pressure case). Such a change in the elastic anisotropy would be characterized by the change of the Zener anisotropy ratio which would become lower than that for pressurized states of YN. It is worth noting that these pressure-induced changes shown in Figure 2b,c are based on zero-pressure second- and third-order elastic constants. In order to check this prediction we have also determined the second-order elastic constants by quantum-mechanical calculations for a series of states at different hydrostatic pressures. Our results are shown in the form of directional dependence of Young's modulus for the hydrostatic pressure of

1.6 GPa in Figure 2d. It can be seen that the softest elastic response (the lowest value of the Young's modulus) is indeed along the $\langle 111 \rangle$ directions.

While we expect that this reversal would be rather rare, we predict it also for ScN with the same rock-salt (B1) structure as in the YN case. Figure 3 visualizes the directional dependence of the Young's modulus for ScN for the zero-pressure case (Figure 3a) as well as the impact of zero-pressure second- and third-order elastic constants on the elasticity of ScN (Figure 3b,c). Performing then calculations of second-order elastic constants also in the case of pressurized ScN (see Figure 3d), the comparison of Figure 3a,d clearly shows the change.

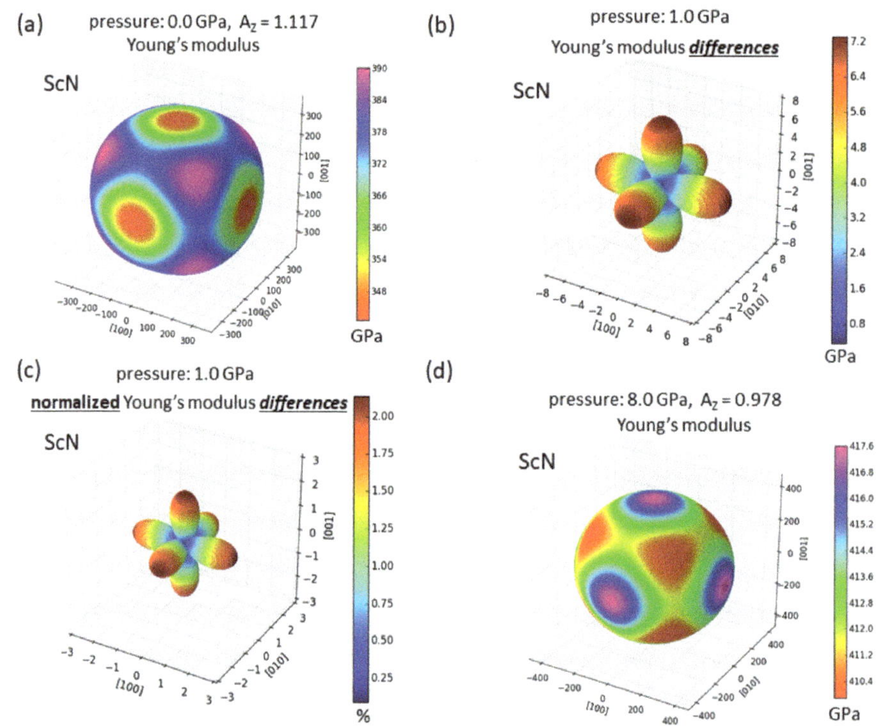

Figure 3. The same as in Figure 2 but for ScN. Part (**d**) is computed at the hydrostatic pressure of 8.0 GPa. The parts (**a**) and (**d**) were visualized by the SC-EMA [42–44] library (scema.mpie.de) based on our ab initio computed elastic constants.

In order to examine the changes in the elasticity in a broader range of pressures we have performed a series of calculations for YN and ScN states with different volumes (different hydrostatic pressures). The resulting elastic constants, the bulk modulus $B = (C_{11} + 2C_{12})/3$, $C' = (C_{11} - C_{12})/2$ and C_{44}, are displayed in Figure 4. As the Zener's ratio could be re-written as $A_Z = C_{44}/C'$ the crossing of the trends for C_{44} and C' indicates the change of the elastic anisotropy type. The pressure dependence of the Zener's ratio is then depicted in Figure 5a. For YN the A_Z ratio reaches the value of 1 (elastic isotropy) for the pressure of about 1.2 GPa. For higher pressures the type of elastic anisotropy is opposite to that in the zero-pressure state. The critical pressure for ScN is about 6.5 GPa (see Figure 5a).

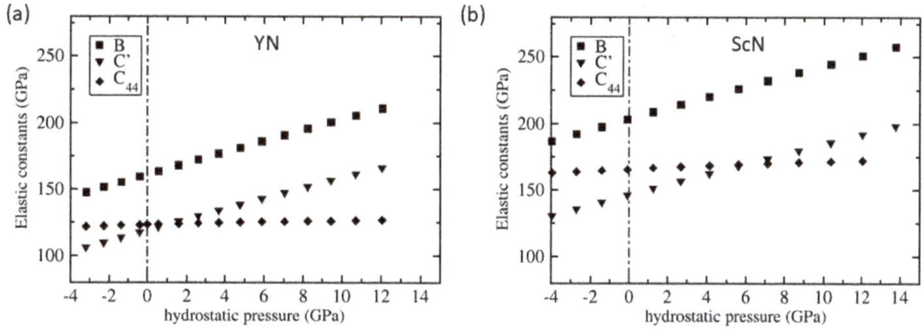

Figure 4. Quantum-mechanically calculated second-order elastic constants of YN (**a**) and ScN (**b**) for different hydrostatic pressures. The vertical dash-dotted lines indicate the zero hydrostatic pressure.

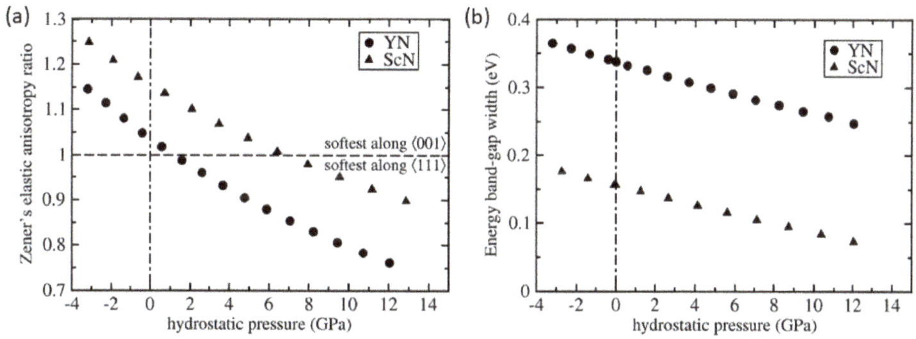

Figure 5. Quantum-mechanically computed (**a**) dependence of the Zener elastic anisotropy ratio A_Z and the band-gap energy (**b**) as a function of hydrostatic pressure for both YN and ScN. The horizontal dashed line for $A_Z = 1$ represents the border value of the elastic anisotropy, the vertical dash-dotted line corresponds to zero hydrostatic pressure.

Next, we have also checked the electronic properties of both materials which are predicted to be semiconductors in their ground state. This is in agreement with previous studies [19,20,23] but our value of the band-gap energy width (0.34 eV) is underestimated similarly as in previous theoretical studies in which similar computational methods were used—see a detailed discussion in Reference [20]. The pressure-dependences of the width of the energy band-gap in their electronic structures are depicted in Figure 5b. It is obvious that both YN and ScN remain semiconducting within the studied range of hydrostatic pressures.

The decreasing width of the energy band-gap with increasing (positive) hydrostatic pressure (lattice constants are smaller than the zero-pressure values) in both materials indicates that there can be a pressure-induced semiconductor-to-metal transition. Such a major change of the electronic structure (and subsequently properties of inter-atomic bonds) may also lead to a phase-instability and a transition into another crystal structure. For example, YN seems to be prone to a pressure-induced phase transition into the B2 (caesium chloride structure) phase according to full potential linearized augmented plane wave (FP-LAPW) calculations in [21,22], but at rather high pressures, 134 GPa [21] and 139 GPa [22]. As none of these transitions seems to be directly related to the change in the type of the elastic anisotropy reported in our present study, we do not examine them in detail and leave them for future work.

As the elastic properties are inter-connected with phonon modes, we have also examined an impact of the above discussed change in the elastic anisotropy on vibrational properties. Our computed phonon spectra for both YN and ScN at the zero-pressure as well at selected pressures (for which the Young's moduli are shown in Figures 2d and 3d) are summarized in Figure 6. The change in the elastic anisotropy is found to have only a very minor impact on the vibrational properties.

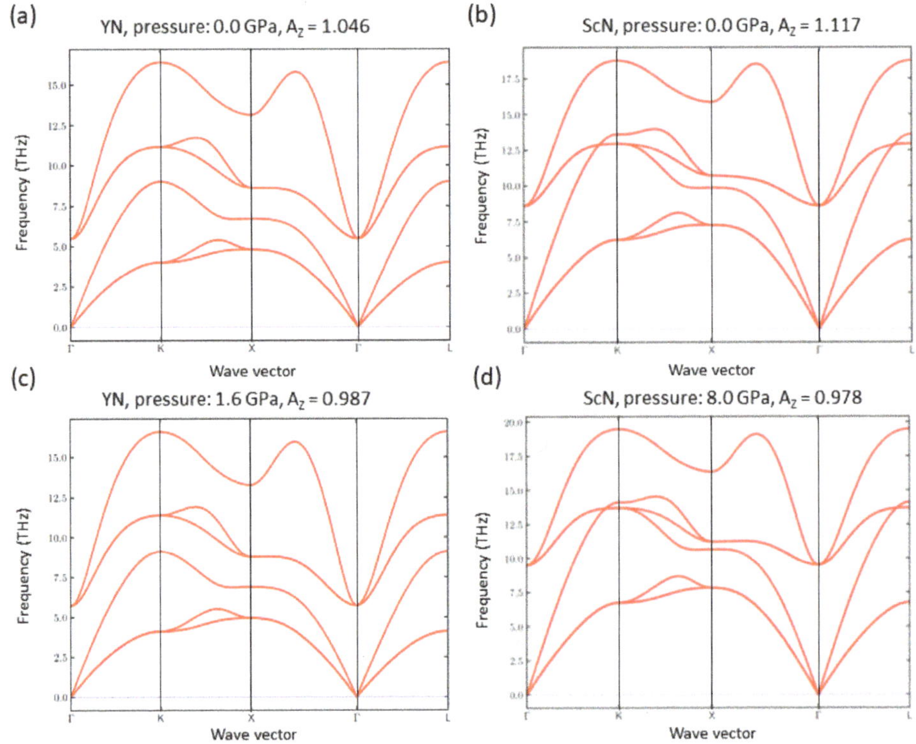

Figure 6. Quantum-mechanically calculated phonon dispersions at zero pressure for YN (**a**) and ScN (**b**) and for YN also for the hydrostatic pressure p = 1.6 GPa (**c**) and for ScN for p = 6.5 GPa (**d**).

After examining in detail the elastic-anisotropy change when applying hydrostatic pressures we next search for other conditions/mechanisms with potentially a similar impact. Our motivation is the fact that hydrostatic pressures over 1 GPa rarely occur in technologically relevant situations. It would be, therefore, desirable to achieve the studied elasticity change under more easily reachable conditions. It is interesting to examine biaxial loading conditions (misfit strains) which are induced, for example, in coherent nanocomposites (such as superlattices [45–66]) when materials with slightly mismatching lattice parameters co-exist. In order to simulate the impact of similar strain conditions, we have performed a series of calculations for tetragonally deformed YN. The YN cell then looses its cubic shape and symmetry and we conveniently describe it by two lattice parameters $a = b$ and c. Considering the fact that these coherently-strained superlattices exist only in the case when the two co-existing materials have their lattice parameters only very slightly different (by about 1–2%), we limit our calculations to $\pm 1.0\%$ change of the lattice parameters with respect to the equilibrium lattice parameter (of the cubic-shape B1 lattice). Figure 7 shows the resulting directional dependencies of the Young's modulus for biaxially 1% compressed case (Figure 7a) and biaxially 1% expanded state (Figure 7b), respectively. The tetragonal lattice parameters c of these states are equal to the values corresponding to the minimum energy (and zero stress $\sigma_c = 0$) with the constraint that the lattice parameters $a = b$ have those specific values.

The state with $a = b = 0.990\ a_{eq}$ and $c = 1.006\ a_{eq}$ (Figure 7a) is characterized by elastic constants $C_{11} = C_{22} = 361$ GPa, $C_{33} = 308$ GPa, $C_{12} = 90$ GPa, $C_{13} = C_{23} = 83$ GPa, $C_{44} = C_{55} = 122$ GPa and $C_{66} = 126$ GPa. The state with $a = b = 1.010\ a_{eq}$ and $c = 0.996\ a_{eq}$ (Figure 7b) has its elasticity described by elastic constants $C_{11} = C_{22} = 283$ GPa, $C_{33} = 325$ GPa, $C_{12} = 75$ GPa, $C_{13} = C_{23} = 79$ GPa, $C_{44} = C_{55} = 123$ GPa and $C_{66} = 120$ GPa.

Importantly, the reversal of the elastic anisotropy is clearly visible in Figure 7. Regarding the biaxially compressed state in Figure 7a, the highest value of the Young's modulus is along the [±100] and [0±10] directions within the (001) plane of the biaxial loading. On the other hand, for the [00±1] directions from the {001} family, the lowest values of the Young's modulus are found. As far as the biaxially expanded state in Figure 7b is concerned, the maximum Young's modulus is along the [00±1] directions perpendicular to the (001) plane of the biaxial loading while the lowest values are obtained for the [±100] and [0±10] directions. The computed differences in the elastic response of the two biaxially loaded states indicate that tetragonal deformations may contribute to fine-tuning of elastic properties within a materials design of systems with a desired elasticity.

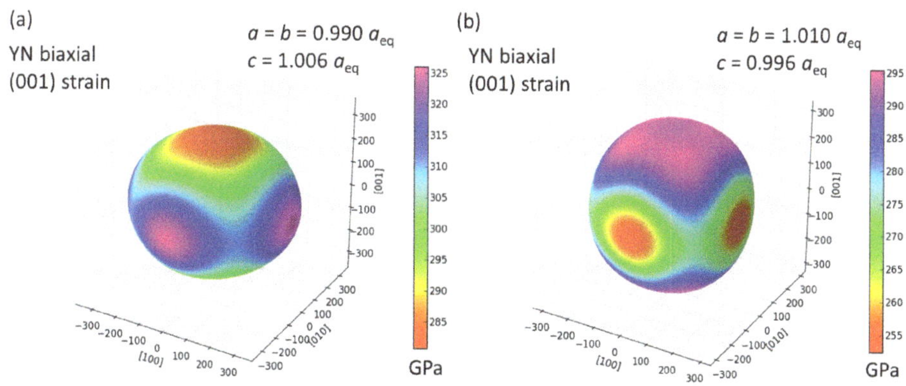

Figure 7. The calculated directional dependences of the Young's modulus of two tetragonally deformed states of YN with $a = b = 0.990\ a_{eq}$ (**a**) $a = b = 1.010\ a_{eq}$ (**b**), respectively. The tetragonal lattice c parameters are equal to the values corresponding to the minimum energy (and zero stress $\sigma_c = 0$).

Our previous simulations of tetragonally-deformed states of YN were motivated by strains appearing in coherent nanocomposites but, admittedly, a hypothetical partner material entered only via the geometry of the imposed strains. In order to check a more realistic situation with two different materials interfacing each other, we next simulate a superlattice consisting of two transition-metal (TM) nitrides each crystalizing in the B1 structure. As it has turned out, it is not easy to find another TM-nitride with the equilibrium lattice parameter close (±1–2%) to that of YN (our value: 4.916 Å). Therefore, we have tried to identify a partner material for ScN for which we found the lattice parameter equal to 4.510 Å. The motivation to replace the hydrostatic pressure of 6.5 GPa by another mechanism is even stronger in the case of ScN. Figure 8a visualizes a 16-atom supercell containing one conventional cell of ScN and one with PdN for which we obtained the lattice parameter equal to 4.447 Å (i.e., 1.4% smaller than in the ScN case). When applying periodic boundary conditions a coherent superlattice is formed. The composite supercell has a tetragonal shape and we will thus use the description by the lattice parameters $a = b$ and c similarly as in the case of tetragonal states of YN discussed above. The calculated values are $a = b = 4.532$ Å and $c = 8.690$ Å. The former value means that ScN is biaxially expanded within the (001) plane by 0.49% and PdN is biaxially expanded by 1.91%. This unexpected results, expansions of both materials, is accompanied by contraction of both materials in the direction [001] perpendicular to the interfaces. In particular, the c lattice parameter of the composite is by 2.98% smaller than the sum of equilibrium lattice parameters of ScN and PdN. The reason for these unexpected results can be probably found in the fact that there is a structural distortion inside the composite. In particular, the N and TM atoms do not share the same planes which are perpendicular to the [001] direction (planes are parallel to the interfaces) as they do in the B1 ground-state structure. The off-sets are alternating (up/down) and their direction are schematically indicated by small arrows in Figure 8a. The magnitude of these shifts is (in relative terms as fractions of the supercell lattice parameter c and as absolute values) equal to ± 0.0061 and ± 0.053 Å in the PdN layers and ± 0.0085 and ± 0.074 Å in ScN layers. Similar shifts of N atoms were found also in MoN/TaN composites [47] and represent quite likely a frozen optical phonon mode.

Figure 8b then shows the elasticity of the studied ScN/PdN nanocomposite. The calculated elastic constants are $C_{11} = C_{22} = 304$ GPa, $C_{33} = 450$ GPa, $C_{12} = 157$ GPa, $C_{13} = C_{23} = 126$ GPa, $C_{44} = C_{55} = 75$ GPa and $C_{66} = 101$ GPa. When inspecting the directional dependence of the Young's modulus in Figure 8b we can observe that the highest values are found for directions close to the [00±1] directions. Thus there is a way how a change of the elastic anisotropy can be achieved also for the ScN. But three aspects should be noted. First, the lowest value of the Young's modulus is along the [±100] and [0±10] directions parallel to the plane of the composite interfaces (see Figure 8b). Second, the overall elastic anisotropy of ScN/PdN nanocomposite is an outcome of complex interactions of pre-strained materials each having a different tensorial elastic properties. Finally, the single-phase PdN in the B1 lattice has, according to our calculations, the Zener's ratio equal to 0.631, i.e., opposite to that of single-phase B1-structure ScN. Therefore, a reduction of the Zener's ratio due to the presence of PdN in the nanocomposite is then expected. Nanocomposites formed by either YN (or ScN) on one hand and other materials on the other require a detailed investigation and will be a topic of future studies.

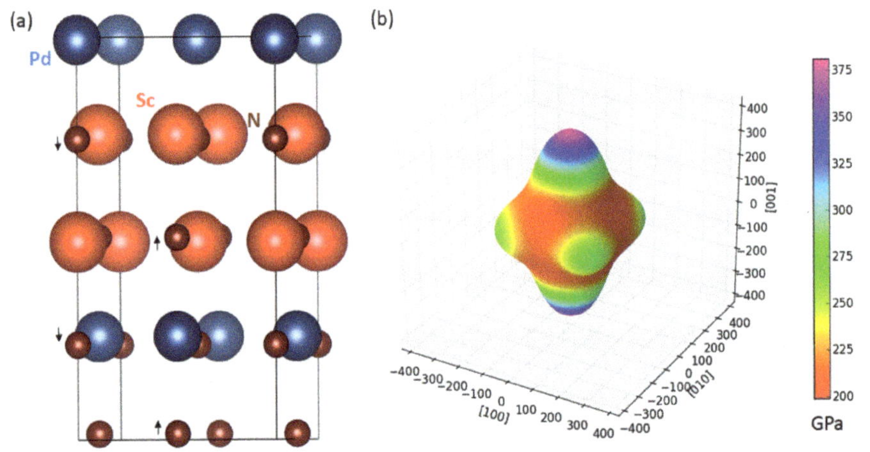

Figure 8. Schematic visualization of 16-atom ScN/PdN supercell (**a**) and the corresponding directional dependence of the Young's modulus (**b**). The calculations for this nanocomposite were performed using $7 \times 7 \times 4$ k-point grid. Small black arrows indicate the shifts of N atoms off the transition-metal planes perpendicular to the [001] direction.

4. Conclusions

We have performed a series of quantum-mechanical calculations of second- and third-order elastic constants of YN and ScN with the rock-salt structure in the case of their zero-pressure states as well as for systems compressed by hydrostatic pressures. We predict that both YN and ScN undergo a reversal of their elastic anisotropy type. In particular, their elastic anisotropy expressed by the

Zener ratio drops under one due to applied hydrostatic pressure. At zero pressure, both systems exhibit the softest elastic response to uniaxial loading (the lowest value of the Young's modulus) along the $\langle 100 \rangle$ directions which is changed to the $\langle 111 \rangle$ directions for pressures beyond the critical one. These transition pressures are rather moderate, 1.2 GPa and 6.5 GPa for YN and ScN, respectively. The elasticity change keeps the semiconducting character of both materials and has only a minor impact on the vibrational properties. As alternative mechanisms leading to the reversal of the elastically soft and hard directions, we identified tetragonal deformations of YN for very small biaxial strains (lattice parameter compressed/expanded by about 1%) and a composite (superlattice) state of ScN and PdN (mismatch of their lattice parameters is 1.4%). The last two discussed mechanisms clearly pave a path towards a strain-controlled fine-tuning of elastic anisotropy in materials, which would allow, for example, a theory-guided design of nanocomposites with a particular ratio of longitudinal sound velocities in the [001], [011] and [111] directions in different components.

Author Contributions: Conceptualization, M.F. and P.K.; Methodology, P.K. and D.H.; Resources, M.Š.; Writing—Original Draft Preparation, M.F.; Writing—Review & Editing, P.K., D.H. and M.Š.; Visualization, P.K. and M.F.; Supervision, M.Š.; Project Administration, M.Š.; Funding Acquisition, M.Š.

Funding: The authors acknowledge the Czech Science Foundation for the financial support received under the Projects Nos. 16-24711S (M.Š.) and 17-18566S (M.F.). Additional resources were provided by the Academy of Sciences of the Czech Republic through the Fellowship of J. E. Purkyně (M.F.) and by the Ministry of Education, Youth and Sports of the Czech Republic under the Project CEITEC 2020, LQ1601 (M.Š). D.H. acknowledges financial support from the Austrian Science Fund (FWF) project Number P30341-N36. The computational results presented have been achieved in part using the Vienna Scientific Cluster (VSC).

Acknowledgments: M.F., P.K. and M.Š. also acknowledge the support from the Academy of Sciences of the Czech Republic (Institutional Project No. RVO:68081723) and from the Ministry of Education, Youth and Sports of the Czech Republic via the research infrastructure IPMINFRA, LM2015069. Computational resources were made available by the Ministry of Education, Youth and Sports of the Czech Republic under the Projects CESNET (Project No. LM2015042), CERIT-Scientific Cloud (Project No. LM2015085) and IT4Innovations National Supercomputer Center (Project No. LM2015070) within the program Projects of Large Research, Development and Innovations Infrastructures. The Figures 1 and 8a were visualized using the VESTA package [67].

Conflicts of Interest: The authors declare no conflict of interest.

References

1. Bacon, D.; Barnett, D.; Scattergood, R. Anisotropic continuum theory of lattice defects. *Prog. Mater. Sci.* **1979**, *23*, 51–262. [CrossRef]
2. Ting, T.C.T. *Anisotropic Elasticity*; Oxford University Press: New York, NY, USA, 1996.
3. Udyansky, A.; von Pezold, J.; Bugaev, V.N.; Friák, M.; Neugebauer, J. Interplay between long-range elastic and short-range chemical interactions in Fe-C martensite formation. *Phys. Rev. B* **2009**, *79*, 224112. [CrossRef]
4. Lothe, J. *Dislocations in Continuous Elastic Media, in Elastic Strain Fields and Dislocation Mobility*; Series of Modern Problems in Condensed Matter Physics; Elsevier: Amsterdam, The Netherlands, 1992; Volume 31.
5. Mouhat, F.; Coudert, F.-X. Necessary and sufficient elastic stability conditions in various crystal systems. *Phys. Rev. B* **2014**, *90*, 224104. [CrossRef]
6. Kraut, E.A. Advances in the theory of anisotropic elastic wave propagation. *Rev. Geophys.* **1963**, *1*, 401–448. [CrossRef]
7. Ting, T.C.T. Longitudinal and transverse waves in anisotropic elastic materials. *Acta Mech.* **2006**, *185*, 147–164. [CrossRef]
8. Thurston, R.; Brugger, K. Third-order elastic constants + velocity of small amplitude elastic waves in homogeneously stressed media. *Phys. Rev.* **1964**, *133*, A1604. [CrossRef]
9. Brugger, K.; Thurston, R. Sound velocity in stressed crystals + 3-order elastic coefficients. *J. Acoust. Soc. Am.* **1964**, *36*, 1041. [CrossRef]
10. Brugger, K. Pure modes for elastic waves in crystals. *J. Appl. Phys.* **1965**, *36*, 759. [CrossRef]
11. Brugger, K. Generalized Gruneisen parameters in anisotropic Debye model. *Phys. Rev.* **1965**, *137*, 1826. [CrossRef]

12. Körmann, F.; Dick, A.; Grabowski, B.; Hallstedt, B.; Hickel, T.; Neugebauer, J. Free energy of bcc iron: Integrated ab initio derivation of vibrational, electronic, and magnetic contributions. *Phys. Rev. B* **2008**, *78*, 033102. [CrossRef]
13. Zhao, J.; Winey, J.M.; Gupta, Y.M. First-principles calculations of second- and third-order elastic constants for single crystals of arbitrary symmetry. *Phys. Rev. B* **2007**, *75*, 094105. [CrossRef]
14. Ledbetter, H.; Naimon, E. Elastic properties of metals and alloys. II. Copper. *J. Phys. Chem. Ref. Data* **1974**, *3*, 897. [CrossRef]
15. Lincoln, R.C.; Koliwad, K.M.; Ghate, P.B. Morse-Potential Evaluation of Second- and Third-Order Elastic Constants of Some Cubic Metals. *Phys. Rev.* **1967**, *157*, 463–466. [CrossRef]
16. De Jong, M.; Winter, I.; Chrzan, D.C.; Asta, M. Ideal strength and ductility in metals from second- and third-order elastic constants. *Phys. Rev. B* **2017**, *96*, 014105. [CrossRef]
17. Kim, K.Y.; Sachse, W.; Every, A.G. On the determination of sound speeds in cubic crystals and isotropic media using a broadband ultrasonic point-source/point-receiver method. *J. Acoust. Soc. Am.* **1993**, *93*, 1393–1406. [CrossRef]
18. Tasnádi, F.; Abrikosov, I.A.; Rogström, L.; Almer, J.; Johansson, M.P.; Odén, M. Significant elastic anisotropy in $Ti_{1-x}Al_xN$ alloys. *Appl. Phys. Lett.* **2010**, *97*, 231902. [CrossRef]
19. Saha, B.; Sands, T.D.; Waghmare, U.V. Electronic structure, vibrational spectrum, and thermal properties of yttrium nitride: A first-principles study. *J. Appl. Phys.* **2011**, *109*, 073720. [CrossRef]
20. Yang, J.W.; An, L. Ab initio calculation of the electronic, mechanical, and thermodynamic properties of yttrium nitride with the rocksalt structure. *Phys. Status Solidi (b)* **2014**, *251*, 792–802. [CrossRef]
21. Mancera, L.; Rodriguez, J.A.; Takeuchi, N. Theoretical study of the stability of wurtzite, zinc-blende, NaCl and CsCl phases in group IIIB and IIIA nitrides. *Phys. Status Solidi (b)* **2004**, *241*, 2424–2428. [CrossRef]
22. Zerroug, S.; Ali Sahraoui, F.; Bouarissa, N. Ab initio calculations of yttrium nitride: Structural and electronic properties. *Appl. Phys. A* **2009**, *97*, 345–350. [CrossRef]
23. Stampfl, C.; Mannstadt, W.; Asahi, R.; Freeman, A.J. Electronic structure and physical properties of early transition metal mononitrides: Density-functional theory LDA, GGA, and screened-exchange LDA FLAPW calculations. *Phys. Rev. B* **2001**, *63*, 155106. [CrossRef]
24. Liu, Z.T.Y.; Zhou, X.; Khare, S.V.; Gall, D. Structural, mechanical and electronic properties of 3d transition metal nitrides in cubic zincblende, rocksalt and cesium chloride structures: A first-principles investigation. *J. Phys. Condens. Matter* **2014**, *26*, 025404. [CrossRef] [PubMed]
25. Mattesini, M.; Magnuson, M.; Tasnádi, F.; Höglund, C.; Abrikosov, I.A.; Hultman, L. Elastic properties and electrostructural correlations in ternary scandium-based cubic inverse perovskites: A first-principles study. *Phys. Rev. B* **2009**, *79*, 125122. [CrossRef]
26. Hohenberg, P.; Kohn, W. Inhomogeneous electron gas. *Phys. Rev.* **1964**, *136*, B864–B871. [CrossRef]
27. Kohn, W.; Sham, L.J. Self-consistent equations including exchange and correlation effects. *Phys. Rev.* **1965**, *140*, A1133–A1138. [CrossRef]
28. Kresse, G.; Hafner, J. Ab initio molecular dynamics for liquid metals. *Phys. Rev. B* **1993**, *47*, 558–561. [CrossRef]
29. Kresse, G.; Furthmüller, J. Efficient iterative schemes for ab initio total-energy calculations using a plane-wave basis set. *Phys. Rev. B* **1996**, *54*, 11169–11186. [CrossRef]
30. Blöchl, P.E. Projector augmented-wave method. *Phys. Rev. B* **1994**, *50*, 17953–17979. [CrossRef]
31. Kresse, G.; Joubert, D. From ultrasoft pseudopotentials to the projector augmented-wave method. *Phys. Rev. B* **1999**, *59*, 1758. [CrossRef]
32. Holec, D.; Friák, M.; Neugebauer, J.; Mayrhofer, P.H. Trends in the elastic response of binary early transition metal nitrides. *Phys. Rev. B* **2012**, *85*, 064101. [CrossRef]
33. Perdew, J.P.; Wang, Y. Accurate and simple analytic representation of the electron-gas correlation energy. *Phys. Rev. B* **1992**, *45*, 13244–13249. [CrossRef]
34. Zhou, L.; Holec, D.; Mayrhofer, P.H. First-principles study of elastic properties of Cr-Al-N. *J. Appl. Phys.* **2013**, *113*, 043511. [CrossRef]
35. Togo, A.; Tanaka, I. First principles phonon calculations in materials science. *Scr. Mater.* **2015**, *108*, 1–5. [CrossRef]
36. Mancera, L.; Rodríguez, J.A.; Takeuchi, N. First principles calculations of the ground state properties and structural phase transformation in YN. *J. Phys. Condens. Matter* **2003**, *15*, 2625. [CrossRef]

37. *Pearson's Handbook of Crystallographic Data for Intermetallic Phases*; American Society for Metals: Metals Park, OH, USA, 1985.
38. Brik, M.; Ma, C.G. First-principles studies of the electronic and elastic properties of metal nitrides XN (X=Sc, Ti, V, Cr, Zr, Nb). *Comput. Mater. Sci.* **2012**, *51*, 380–388. [CrossRef]
39. Gall, D.; Petrov, I.; Hellgren, N.; Hultman, L.; Sundgren, J.E.; Greene, J.E. Growth of poly- and single-crystal ScN on MgO(001): Role of low-energy N_2^+ irradiation in determining texture, microstructure evolution, and mechanical properties. *J. Appl. Phys.* **1998**, *84*, 6034–6041. [CrossRef]
40. Birch, F. Finite Elastic Strain of Cubic Crystals. *Phys. Rev.* **1947**, *71*, 809. [CrossRef]
41. Perdew, J.P.; Burke, K.; Ernzerhof, M. Generalized Gradient Approximation Made Simple. *Phys. Rev. Lett.* **1996**, *77*, 3865–3868. [CrossRef] [PubMed]
42. Titrian, H.; Aydin, U.; Friák, M.; Ma, D.; Raabe, D.; Neugebauer, J. Self-consistent Scale-bridging Approach to Compute the Elasticity of Multi-phase Polycrystalline Materials. *MRS Proc.* **2013**, *1524*, mrsf12-1524-rr06-03. [CrossRef]
43. Friák, M.; Counts, W.; Ma, D.; Sander, B.; Holec, D.; Raabe, D.; Neugebauer, J. Theory-Guided Materials Design of Multi-Phase Ti-Nb Alloys with Bone-Matching Elastic Properties. *Materials* **2012**, *5*, 1853–1872. [CrossRef]
44. Zhu, L.F.; Friák, M.; Lymperakis, L.; Titrian, H.; Aydin, U.; Janus, A.; Fabritius, H.O.; Ziegler, A.; Nikolov, S.; Hemzalová, P.; Raabe, D.; Neugebauer, J. Ab initio study of single-crystalline and polycrystalline elastic properties of Mg-substituted calcite crystals. *J. Mech. Behav. Biomed. Mater.* **2013**, *20*, 296–304. [CrossRef] [PubMed]
45. Mayrhofer, P.H.; Fischer, F.D.; Boehm, H.J.; Mitterer, C.; Schneider, J.M. Energetic balance and kinetics for the decomposition of supersaturated Ti1-xAlxN. *Acta Mater.* **2007**, *55*, 1441–1446. [CrossRef]
46. Wu, L.; Chen, M.; Li, C.; Zhou, J.; Shen, L.; Wang, Y.; Zhong, Z.; Feng, M.; Zhang, Y.; Han, K.; et al. Ferromagnetism and matrix-dependent charge transfer in strained $LaMnO_3$-$LaCoO_3$ superlattices. *Mater. Res. Lett.* **2018**, *6*, 501–507. [CrossRef]
47. Koutná, N.; Holec, D.; Friák, M.; Mayrhofer, P.H.; Šob, M. Stability and elasticity of metastable solid solutions and superlattices in the MoN–TaN system: First-principles calculations. *Mater. Des.* **2018**, *144*, 310–322. [CrossRef]
48. Jiang, M.; Xiao, H.Y.; Peng, S.M.; Yang, G.X.; Liu, Z.J.; Zu, X.T. A comparative study of low energy radiation response of AlAs, GaAs and GaAs/AlAs superlattice and the damage effects on their electronic structures. *Sci. Rep.* **2018**, *8*. [CrossRef] [PubMed]
49. Wen, Y.N.; Gao, P.F.; Xia, M.G.; Zhang, S.L. Half-metallic ferromagnetism prediction in MoS2-based two-dimensional superlattice from first-principles. *Mod. Phys. Lett. B* **2018**, *32*. [CrossRef]
50. Friák, M.; Tytko, D.; Holec, D.; Choi, P.P.; Eisenlohr, P.; Raabe, D.; Neugebauer, J. Synergy of atom-probe structural data and quantum-mechanical calculations in a theory-guided design of extreme-stiffness superlattices containing metastable phases. *New J. Phys.* **2015**, *17*, 093004. [CrossRef]
51. Dai, Q.; Eckern, U.; Schwingenschlog, U. Effects of oxygen vacancies on the electronic structure of the $(LaVO_3)_6/SrVO_3$ superlattice: A computational study. *New J. Phys.* **2018**, *20*. [CrossRef]
52. Jiang, M.; Xiao, H.; Peng, S.; Qiao, L.; Yang, G.; Liu, Z.; Zu, X. First-Principles Study of Point Defects in GaAs/AlAs Superlattice: the Phase Stability and the Effects on the Band Structure and Carrier Mobility. *Nanoscale Res. Lett.* **2018**, *13*. [CrossRef]
53. Chen, H.; Millis, A.J.; Marianetti, C.A. Engineering Correlation Effects via Artificially Designed Oxide Superlattices. *Phys. Rev. Lett.* **2013**, *111*. [CrossRef]
54. Mottura, A.; Janotti, A.; Pollock, T.M. A first-principles study of the effect of Ta on the superlattice intrinsic stacking fault energy of $L1_2$-Co_3(Al,W). *Intermetallics* **2012**, *28*, 138–143. [CrossRef]
55. Rosengaard, N.; Skriver, H. Ab-initio study of antiphase boundaries and stacking-faults in $L1_2$ and $D0_{22}$ compounds. *Phys. Rev. B* **1994**, *50*, 4848–4858. [CrossRef]
56. Torres-Pardo, A.; Gloter, A.; Zubko, P.; Jecklin, N.; Lichtensteiger, C.; Colliex, C.; Triscone, J.M.; Stephan, O. Spectroscopic mapping of local structural distortions in ferroelectric $PbTiO_3/SrTiO_3$ superlattices at the unit-cell scale. *Phys. Rev. B* **2011**, *84*. [CrossRef]
57. Chawla, V.; Holec, D.; Mayrhofer, P.H. Stabilization criteria for cubic AlN in TiN/AlN and CrN/AlN bi-layer systems. *J. Phys. D Appl. Phys.* **2013**, *46*. [CrossRef]

58. Cooper, V.R.; Rabe, K.M. Enhancing piezoelectricity through polarization-strain coupling in ferroelectric superlattices. *Phys. Rev. B* **2009**, *79*. [CrossRef]
59. Chen, B.; Zhang, Q.; Bernholc, J. Si diffusion in gaas and si-induced interdiffusion in gaas/alas superlattices. *Phys. Rev. B* **1994**, *49*, 2985–2988. [CrossRef]
60. Schmid, U.; Christensen, N.; Cardona, M.; Lukes, F.; Ploog, K. Optical anisotropy in GaAs/AlSs(110) superlattices. *Phys. Rev. B* **1992**, *45*, 3546–3551. [CrossRef]
61. Gibson, Q.D.; Schoop, L.M.; Weber, A.P.; Ji, H.; Nadj-Perge, S.; Drozdov, I.K.; Beidenkopf, H.; Sadowski, J.T.; Fedorov, A.; Yazdani, A.; et al. Termination-dependent topological surface states of the natural superlattice phase Bi_4Se_3. *Phys. Rev. B* **2013**, *88*. [CrossRef]
62. Park, C.; Chang, K. Structural and electronic-properties of gap-alp (001) superlattices. *Phys. Rev. B* **1993**, *47*, 12709–12715. [CrossRef]
63. Romanyuk, O.; Hannappel, T.; Grosse, F. Atomic and electronic structure of GaP/Si(111), GaP/Si(110), and GaP/Si(113) interfaces and superlattices studied by density functional theory. *Phys. Rev. B* **2013**, *88*. [CrossRef]
64. Abdulsattar, M.A. SiGe superlattice nanocrystal pure and doped with substitutional phosphorus single atom: Density functional theory study. *Superlattices Microstruct.* **2011**, *50*, 377–385. [CrossRef]
65. Botti, S.; Vast, N.; Reining, L.; Olevano, V.; Andreani, L. Ab initio and semiempirical dielectric response of superlattices. *Phys. Rev. B* **2004**, *70*. [CrossRef]
66. Rondinelli, J.M.; Spaldin, N.A. Electron-lattice instabilities suppress cuprate-like electronic structures in $SrFeO_3/OSrTiO_3$ superlattices. *Phys. Rev. B* **2010**, *81*. [CrossRef]
67. Momma, K.; Izumi, F. VESTA 3 for three-dimensional visualization of crystal, volumetric and morphology data. *J. Appl. Crystallogr.* **2011**, *44*, 1272–1276. [CrossRef]

© 2018 by the authors. Licensee MDPI, Basel, Switzerland. This article is an open access article distributed under the terms and conditions of the Creative Commons Attribution (CC BY) license (http://creativecommons.org/licenses/by/4.0/).

nanomaterials

Article

Graphene Surface Reinforcement of Iron

Pengjie Wang [1], Qiang Cao [1,*], Yuping Yan [2], Yangtian Nie [1], Sheng Liu [2] and Qing Peng [1,3,*]

1 The Institute of Technological Sciences, Wuhan University, Wuhan 430072, China; pjwang@whu.edu.cn (P.W.); nieyangtian@whu.edu.cn (Y.N.)
2 School of Power and Mechanical Engineering, Wuhan University, Wuhan 430072, China; yypgoodluck87@163.com (Y.Y.); victor_liu63@126.com (S.L.)
3 Nuclear Engineering and Radiological Sciences, University of Michigan, Ann Arbor, MI 48109, USA
* Correspondence: caoqiang@whu.edu.cn (Q.C.); qpeng.org@gmail.com (Q.P.); Tel.: +86-137-0129-2834 (Q.C.); +1-734-763-3866 (Q.P.)

Received: 14 December 2018; Accepted: 28 December 2018; Published: 4 January 2019

Abstract: Graphene is an ideal material in the reinforcement of metal-matrix composites owing to its outstanding mechanical and physical properties. Herein, we have investigated the surface enhancement of iron via a computational nanoindentation process using molecular dynamics simulations. The findings of our study show that graphene can enhance the critical yield strength, hardness and elastic modulus of the composite to different degrees with the change of the number of graphene layers. In the six tested models, the composite with trilayer graphene on the surface produces the strongest reinforcement, with an increased magnitude of 432.1% and 169.5% in the hardness and elastic modulus, respectively, compared with pure iron. Furthermore, it is revealed that high temperature could weaken the elastic bearing capacity of the graphene, resulting in a decrease on the elastic mechanical properties of the graphene/Fe composite.

Keywords: nanoindentation; graphene/Fe composite; critical yield strength; hardness; elastic modulus

1. Introduction

Since its discovery [1], graphene has been the most attractive material to be explored because of its remarkable electronic and physical properties due to the quantum confinement [2–5]. In particular, graphene has shown great potential in matrix reinforcement in recent years owing to its outstanding strength over 1 TPa [5]. In general, a low modulus matrix can be significantly reinforced by the presence of high-modulus graphene, which is called filler in composites [6]. The common assumption that the filler modulus is independent of the matrix has been proved incorrect, considering the wide range of reinforcement on polymer matrices by high-modulus graphene [6]. Many factors influence the mechanical properties of graphene-based nanocomposites, including the structure of the filler, the synthetic method of the composite, the concentration of the filler in the matrix, the interactions between the filler and the matrix, and the orientation of the filler. Even with a very small amount of graphene, the composite's Young's modulus, tensile strength and toughness can have sharp increases [7,8].

Extensive studies have been conducted on metal matrix composites, including Fe, Al, Cu, Mg and Ni [9–13]. Iron is by far the most commonly used industrial metal on account of its great range of desirable properties and low cost. Dislocations play an important role in revealing the remarkable mechanical properties of iron matrix composite, about which a punch-out mechanism has been proposed to explain the formation of interstitial dislocation loops [14]. The study on the interaction between edge dislocations and graphene nanosheets in graphene/Fe composites by molecular dynamics (MD) simulations revealed an increase of 107% and 1400% in shear modulus and yield stress, respectively [9]. Meanwhile, the enhancement of surface hardness on iron is another crucial issue in broad industrial applications of graphene. Graphene is an excellent choice for the surface reinforcement of pure iron matrix due to the relatively simple industrial process, the study on which can offer guidance to the manufacture and application of the graphene/Fe composite. However, relevant efforts have been rarely reported.

Nanoindentation is an approach widely used to measure elastic modulus and hardness of nanocomposites [15,16]. In this study we investigated the enhancement of hardness and elastic modulus of iron by graphene additives. The amount of graphene measured by the number of layers was explicitly examined using nanoindentation modeling. We recorded the load-displacement data during the nanoindentation process of the graphene/Fe composite made by a diamond indenter. By analyzing the load-displacement curves, we compared the hardness and elastic modulus between the pure iron matrix and composites with graphene on the surface and in the superficial zone. In addition, the influence of different loading speeds and temperatures on the elastic mechanical properties of the composite were also discussed.

2. Method

2.1. Model

The MD method was used to examine the influence of graphene in the graphene/Fe composite at the atomic level. The large-scale atomic/molecular massively parallel simulator (LAMMPS) software was employed to calculate the MD simulations. The simulation cells are shown in Figure 1. The x and y directions are set the periodic boundary condition and the z direction is fixed.

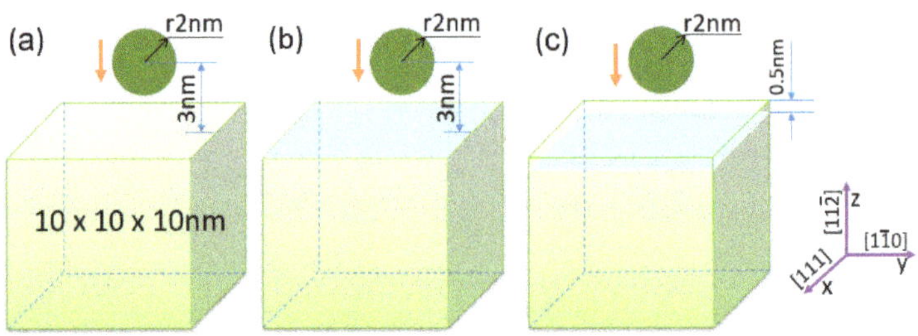

Figure 1. (**a**) Simulation cell of pure iron matrix; (**b**) simulation cell of composite with graphene on the surface of the matrix, the graphene could be monolayer, bilayer and trilayer; (**c**) simulation cell of composite with graphene in the superficial zone of the matrix, the graphene could be monolayer, bilayer and trilayer.

The graphene/Fe composite models are shown in Figure 1b,c. In this study the BCC α-Fe matrix is an area of 10 × 10 × 10 nm with a lattice constant of 2.85 Å, which is shown in Figure 1a. Figure 1b shows the composite model with graphene on the surface of the matrix. The composite with graphene in the superficial zone of the matrix is shown in Figure 1c, where the distance between the graphene layer and the top surface is 0.5 nm. All balls with a radius of 2 nm consisting of carbon atoms in diamond structure have the same velocity moving down to make a nanoindentation on the surface of the composite or pure iron matrix.

2.2. Molecular Dynamics Simulations

The accuracy of the MD simulation results is determined by the selection of potential function. We used the Brenner-generation reactive empirical bond-order potential to model the C-C bonded interaction [17]. The C-C AIREBO potential has a widespread application in graphene-based materials [18,19], which is composed of three terms:

$$\mathrm{E} = \frac{1}{2}\sum_{i}\sum_{j\neq i}\left[E_{ij}^{REBO} + E_{ij}^{LJ} + \sum_{k\neq i,j}\sum_{l\neq i,j,k}E_{kijl}^{TORSION}\right], \tag{1}$$

where the E_{ij}^{REBO} term describes the short-ranged interactions (r < 2 Å) between carbon atoms, the E_{ij}^{LJ} term adds longer-ranged interactions (2 Å < r < cutoff) using a form similar to the standard Lennard Jones potential, and the $E_{kijl}^{TORSION}$ term describes various dihedral angle preferences in hydrocarbon configurations, which is an explicit 4-body potential.

The embedded-atom method (EAM) potential was used to compute pairwise interactions between iron atoms [20]. The total energy E_i of an atom i is specified as

$$E_i = F_\alpha\left(\sum_{j\neq i}\rho_\beta(r_{ij})\right) + \frac{1}{2}\sum_{j\neq i}\varnothing_{\alpha\beta}(r_{ij}), \tag{2}$$

where F_α is the embedding energy which is a function of the atomic electron density ρ_β. $\Phi_{\alpha\beta}$ is the pair potential interaction between atoms I and J, as a function of the distance r_{ij} between atom I and atom J. α and β are atomic element types. The C-Fe interaction between iron and carbon atoms of both diamond indenter and graphene layers was described by the classical Lennard Jones (LJ) potential, as shown in Equation (3)

$$\mathrm{E} = 4\varepsilon\left[\left(\frac{\sigma}{r}\right)^{12} - \left(\frac{\sigma}{r}\right)^{12}\right]\ r < r_c, \tag{3}$$

where r_c is the cutoff distance. The σ and ε for the C-Fe interaction are 2.221 Å and 0.043 eV [21], respectively. The LJ potential was also used to model the interaction between the graphene layers and diamond indenter with the σ and ε value of 3.4 Å and 0.00284 eV [22], respectively.

We set a downward velocity of 30 m/s on the balls and kept the temperature at 300 K by the Nose-Hoover algorithm [23]. After the energy minimization, the downward movement would continue until a preset depth and then the ball moved upward. The timestep was 0.001 ps. The displacement and the force in the z-direction of the diamond indenter imposed by the composite matrix were recorded.

2.3. Nanoindentation

The Hertzian contact analysis theory guided our model. To understand the influence of graphene sheets on the hardness and elastic modulus of the iron matrix, the load-displacement data was recorded for analysis. The load-displacement relation is described by [24]

$$\mathrm{P} = \frac{4}{3}\sqrt{R}E_r\left(h - h_f\right)^{\frac{3}{2}}, \tag{4}$$

where P is the z-direction force of diamond indenter, h is the max depth of the nanoindentation, h_f is the final displacement of the plastic unloading process. R is defined by $R = (1/R_1 + 1/R_2)^{-1}$, where R_1 is the radius of diamond indenter, R_2 is the radius of the spherical hole in the surface of the substrate [24]. The induced modulus E_r is given by [25]

$$\frac{1}{E_r} = \frac{(1-\nu^2)}{E} + \frac{(1-\nu_i^2)}{E_i}, \tag{5}$$

where E and ν are Elastic modulus (Young's modulus) and Poisson's ratio of the composite matrix, E_i and ν_i are the same parameters of the diamond indenter. The hardness is defined by

$$\mathrm{H} = \frac{P_{max}}{A_c}, \tag{6}$$

where P_{max} is the max force at the initial unloading point, and A_c is the contact area of spherically curved surface [25].

3. Results

3.1. Monolayer Graphene Enhancement

The results of the simulated nanoindentation process are revealed by the load-displacement curves. Figure 2a shows the load-displacement curves of the three cases, i.e., pure iron, monolayer graphene on the surface and monolayer graphene in the superficial zone of the graphene/Fe composite. The max displacement of the diamond indenter is 4.5 nm. These results indicate that the monolayer graphene has a significant reinforcement on iron matrix both in surface and superficial zone cases. In the initial 1 nm displacement, the load of the three cases shares almost the same growth rate, indicating that graphene has no evident effect on iron matrix in the early stage of loading. After that, the pure iron loading curve gradually levels off, while the curves of the other two composites keep growing. The difference in curves indicates that graphene acts as a strong deterrent for the yield of the composite. From Figure 2a, we can see that the graphene has increased the yield strength of the composites obviously. The sharp drop points of the curves indicate that the load bearing capability of the graphene has reached its maximum and then the graphene is fracted. After that, the load curves of the three cases begin to converge. We made the loading-unloading processes of those three cases in their plastic deformation stages. The load change of the three cases is shown in Figure 2b. The arrows indicate the loading and unloading processes. The serration in the curve of pure iron is mainly related to the potential we adopted to describe the C-Fe interaction. The elastic modulus and hardness of pure iron and the two composite cases with monolayer graphene were calculated according to the loading-unloading curves. The hardness of pure iron is 8.1, very close to the experimental value of 8.2 [26]. The elastic modulus of pure iron is 150.1 GPa, a little lower than the experimental value of 200 GPa [26] and a MD result in shear modulus of 56.4 GPa [9], which is associated with the potentials used in the simulation. The two parameters of the composite with monolayer graphene on the surface are 19.4 and 218.4 GPa, an increase of 139.5% and 45.5%, respectively, compared with the pure iron case. In the case of composites with graphene in the superficial zone, the hardness and elastic modulus are 18.4 and 201.1 GPa, an increase of 127.2% and 33.3%, respectively. In general, the monolayer graphene has greatly increased the hardness and elastic modulus of the composite due to its remarkable load bearing capability. Meanwhile, the composite with graphene on the surface produces a better reinforcement. The local stress and interspace would be produced as the graphene embedded into the iron matrix, which enhances the flexural rigidity and weakens the elastic bearing capacity of the graphene, resulting in a decrease of the hardness and elastic modulus.

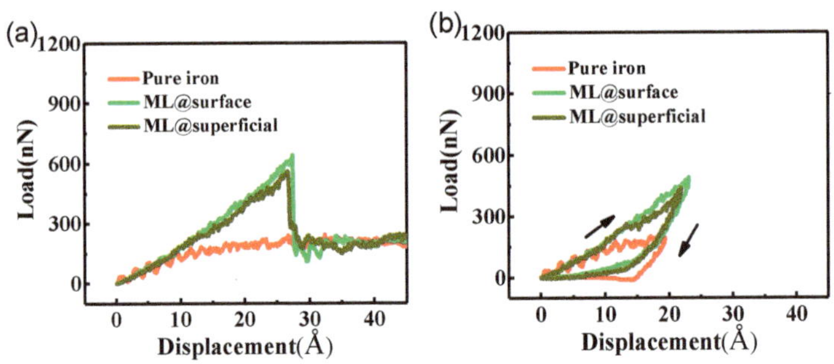

Figure 2. *Cont.*

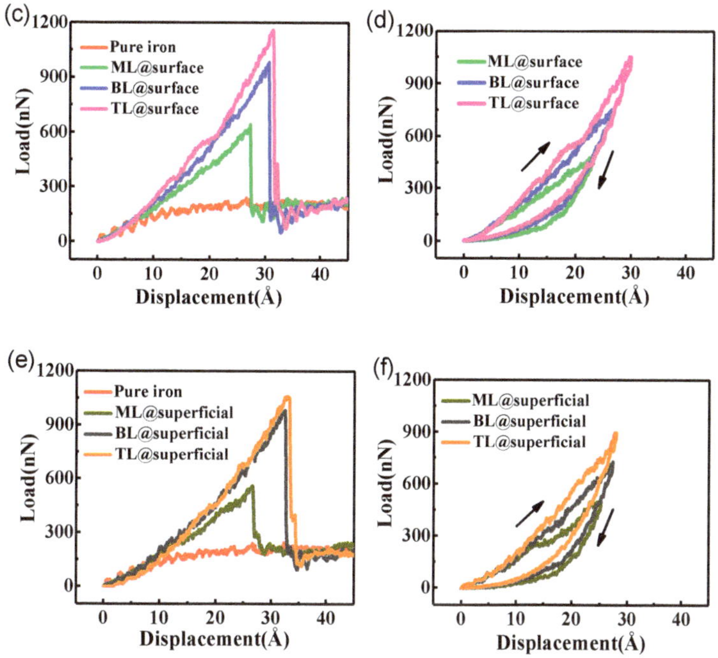

Figure 2. (**a**) Load-displacement curves of pure iron, composite with monolayer (ML) graphene on surface and in superficial zone; (**b**) loading-unloading curves of pure iron and composite with monolayer graphene; (**c**) load-displacement curves of pure iron and composite with different numbers of layers on the surface (ML, bilayer (BL), trilayer (TL)); (**d**) loading-unloading curves of three surface composite cases; (**e**) load-displacement curves of pure iron and composite with different number layers in superficial zone; (**f**) loading-unloading curves of three superficial composite cases.

3.2. Multilayer Graphene Enhancement

The mechanical properties of multiple graphene have received growing interest in recent years [27,28]. To investigate the reinforcement of graphene/Fe composites with different numbers of graphene layers, we explored monolayer, bilayer and trilayer graphene on the surface and in the superficial zone. Figure 2c shows the loading processes of the composite with different numbers of graphene layers on the surface. It was observed that the critical yield strength of the composite increases with the number of graphene layers, due to the improvement of the load bearing capability by adding more graphene layers. The critical nanoindentation depth is also extended from 2.7 to 3.2 nm, owing to the enhancement of elastic deformation capacity of the composite, as the graphene increases from monolayer to trilayer. Figure 2d shows the loading-unloading curves of the three surface cases. The hardness and elastic modulus of composite with bilayer graphene on the surface shown in Figure 3 are 28.9 and 286.5 GPa, with increases of 256.8% and 90.9%, respectively, compared with the pure iron case. The same parameters of composite with trilayer graphene on the surface are 43.1 and 404.5 GPa, corresponding to increases of 432.1% and 169.5%, respectively, compared with the pure iron case. The results suggest that the graphene layers have an effective improvement on the elastic mechanical properties of the iron matrix. With the increase of graphene layers, the max load increases significantly while the contact area changes slightly, leading to the proportional increase of hardness and elastic modulus. Figure 2e shows the loading processes of the composites with graphene in the superficial zone. The relationship between yield strength and graphene layers is similar to the surface cases. However, the max yield strength of bilayer graphene case is only a little smaller than the trilayer case in the superficial zone cases. In the superficial trilayer case, the local stress was produced on the top layer and under the bottom layer graphene region, which caused the rugged surfaces under the top graphene layer and on the bottom graphene layer. The middle layer graphene appeared to slip due to the unbalanced van der Waals force from the other two graphene layers at the initial time. The degree of distortion on the middle graphene layer is uneven during the nanoindentation process, which weakens the loading bearing capacity in a way. As a result, the max yield strength of trilayer case did not improve much more than the bilayer case. The loading-unloading curves of the three cases in superficial zone are shown in Figure 2f. Figure 3 also shows the hardness and elastic modulus of composite with different number layers graphene in superficial zone. The hardness and elastic modulus are 24.8 and 275.0 GPa for the bilayer case, 29.6 and 347.1 GPa for the trilayer case. The increments of those two parameters are 206.2% and 83.2% for bilayer case, 265.4% and 131.2% for the trilayer case, compared with the pure iron case. The variation tendency on elastic modulus and hardness of all the surface and superficial cases are shown in Figure 3a,b, respectively. The two parameters of composite in surface cases are higher than that in superficial zone cases to different degrees with the same number of graphene layers, which is related to the local stress fields produced by lattice mismatch between the graphene and iron matrix in the superficial situations. The stress fields reduce the flexibility of graphene layers and lower the critical yield strength of the graphene/Fe composite.

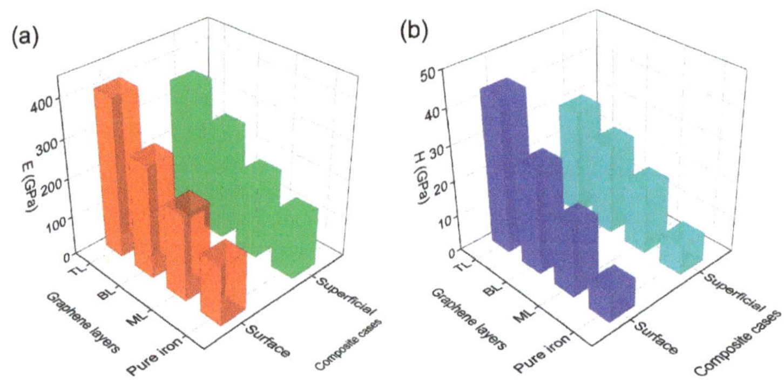

Figure 3. (**a**) Elastic modulus of composite cases (pure iron, ML, BL, TL); (**b**) hardness of composite cases.

3.3. *Effect of Loading Speed and Temperature*

To investigate the influence of loading speed and temperature on the nanoindentation process of the graphene/Fe composite, we explored the nanoindentation simulation of the composite with monolayer graphene on surface at varied loading speeds and at different temperatures, as shown in Figure 4a,b, respectively. It can be seen that the critical yield strength in the case with a loading speed of 100 m/s is a little higher than the other two cases, which suggests that the change of loading speed has a slight effect on the loading bearing capacity of the graphene. The situation for 10 m/s resembles the 30 m/s case, and all the speed cases have almost the same critical nanoindentation depth, which reveals that the elastic bearing capacity of the composite has no relationship with the loading speed. Meanwhile, the effect of different temperatures is more obvious. The critical yield strength and nanoindentation depth decreased significantly with the increase of temperature, which is related to the effect on the crystal texture of the iron matrix and graphene by different temperatures. As the temperature increases, the lattice vibration of graphene becomes intensified, which greatly reduces its mechanical properties, resulting in a weakening of elastic bearing capacity [29]. Meanwhile, the high temperature improves the atomic activity, lowers the grain boundary strength and reduces the lattice resistance in the iron matrix region, which also has a negative effect on the loading bearing capacity of the composite.

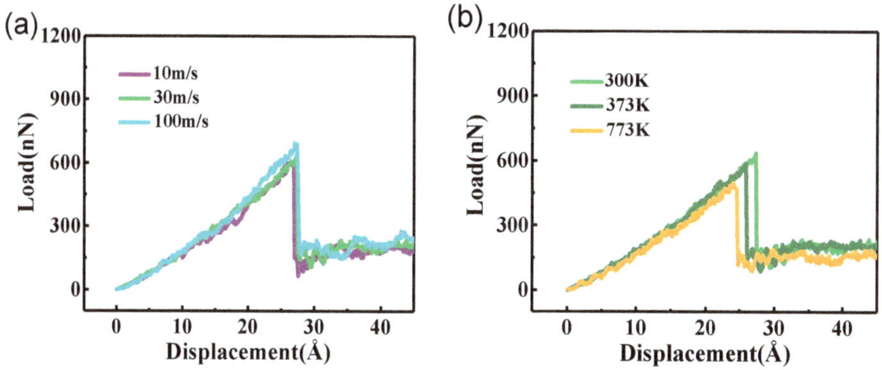

Figure 4. (**a**) Load-displacement curves of composite with monolayer graphene on surface in different loading speeds; (**b**) load-displacement curves of composite with monolayer graphene on surface at different temperatures.

The simulation results reveal the change of hardness and elastic modulus of iron matrix composites with the graphene of the different number of layers, which is well consistent with previous studies on metal-matrix composites reinforced by graphene [30,31]. Both the surface and the superficial composite cases have a significant reinforcement on the iron matrix to different degrees. In our models, the composite with trilayer graphene on the surface has the strongest hardness of 43.1 GPa and the maximum elastic modulus of 404.5 GPa, corresponding to an increase of 432.1% and 169.5%, respectively, compared to the pure iron case. Further studies could be conducted to optimize the results, such as the size of the diamond indenter, potential functions and the system size. Furthermore, the reinforcement caused by the graphene in other morphologies in the iron matrix is also expected to be further explored.

4. Conclusions

To conclude, we have investigated the mechanical properties change of graphene/Fe composites with different number of graphene layers on the surface and in the superficial zone by using the MD simulation method to model the nanoindentation process. The results from the loading-displacement curves show that the graphene layers have a significant effect on the critical yield strength, hardness and elastic modulus of iron matrix in different degrees due to its remarkable loading bearing capacity. The degree of reinforcement on iron matrix increases with the number of graphene layers, both in surface and superficial zone cases. It was demonstrated that the enhancement in hardness and elastic modulus of composite in the surface cases is slightly stronger than that in superficial zone cases with graphene of the same number of layers on account of the production of the local stress fields in the superficial situations. Finally, the loading speeds have a small effect on the simulation of the composite nanoindentation process, and the simulation results of different temperatures reveal that high temperature can effectively reduce the mechanical properties of the graphene/Fe composite by softening lattice rigidity.

Author Contributions: P.W. conceived the idea of the paper. P.W., Q.C. and Q.P. wrote the paper. Y.Y., Y.N. and S.L. offer many helpful advices. All the authors had a full discussion and comment on the paper.

Funding: We acknowledge the calculation support of the Supercomputing Center of Wuhan University and the financial support from the National Natural Science Foundation of China (No. 51727901).

Acknowledgments: We want to acknowledge Pinxuan He, Xiao Geng, Huaipeng Wang for their discussions and helpful support. The numerical calculations in this paper were done on the supercomputing system in the Supercomputing Center of Wuhan University.

Conflicts of Interest: The authors declare no conflict of interest.

References

1. Novoselov, K.S.; Geim, A.K.; Morozov, S.V.; Jiang, D.; Zhang, Y.; Dubonos, S.V.; Grigorieva, I.V.; Firsov, A.A. Electric field effect in atomically thin carbon films. *Science* **2004**, *306*, 666–669. [CrossRef] [PubMed]
2. Tombros, N.; Jozsa, C.; Popinciuc, M.; Jonkman, H.T.; van Wees, B.J. Electronic spin transport and spin precession in single graphene layers at room temperature. *Nature* **2007**, *448*, 571–574. [CrossRef] [PubMed]
3. Bunch, J.S.; van der Zande, A.M.; Verbridge, S.S.; Frank, I.W.; Tanenbaum, D.M.; Parpia, J.M.; Craighead, H.G.; McEuen, P.L. Electromechanical resonators from graphene sheets. *Science* **2007**, *315*, 490–493. [CrossRef] [PubMed]
4. Novoselov, K.S. Nobel Lecture: Graphene: Materials in the Flatland. *Rev. Mod. Phys.* **2011**, *83*, 837–849. [CrossRef]
5. Lee, C.; Wei, X.D.; Kysar, J.W.; Hone, J. Measurement of the elastic properties and intrinsic strength of monolayer graphene. *Science* **2008**, *321*, 385–388. [CrossRef] [PubMed]
6. Papageorgiou, D.G.; Kinloch, I.A.; Young, R.J. Mechanical properties of graphene and graphene-based nanocomposites. *Prog. Mater. Sci.* **2017**, *90*, 75–127. [CrossRef]
7. Papageorgiou, D.G.; Kinloch, I.A.; Young, R.J. Graphene/elastomer nanocomposites. *Carbon* **2015**, *95*, 460–484. [CrossRef]
8. Naebe, M.; Wang, J.; Amini, A.; Khayyam, H.; Hameed, N.; Li, L.H.; Chen, Y.; Fox, B. Mechanical Property and Structure of Covalent Functionalised Graphene/Epoxy Nanocomposites. *Sci. Rep.* **2014**, *4*, 4375. [CrossRef]
9. Wang, L.; Jin, J.F.; Cao, J.Y.; Yang, P.J.; Peng, Q. Interaction of Edge Dislocations with Graphene Nanosheets in Graphene/Fe Composites. *Crystals* **2018**, *8*, 160. [CrossRef]
10. Bartolucci, S.F.; Paras, J.; Rafiee, M.A.; Rafiee, J.; Lee, S.; Kapoor, D.; Koratkar, N. Graphene-aluminum nanocomposites. *Mater. Sci. Eng. a-Struct.* **2011**, *528*, 7933–7937. [CrossRef]
11. Hwang, J.; Yoon, T.; Jin, S.H.; Lee, J.; Kim, T.S.; Hong, S.H.; Jeon, S. Enhanced Mechanical Properties of Graphene/Copper Nanocomposites Using a Molecular-Level Mixing Process. *Adv. Mater.* **2013**, *25*, 6724–6729. [CrossRef]
12. Chen, L.Y.; Konishi, H.; Fehrenbacher, A.; Ma, C.; Xu, J.Q.; Choi, H.; Xu, H.F.; Pfefferkorn, F.E.; Li, X.C. Novel nanoprocessing route for bulk graphene nanoplatelets reinforced metal matrix nanocomposites. *Scripta Mater.* **2012**, *67*, 29–32. [CrossRef]
13. Kim, Y.; Lee, J.; Yeom, M.S.; Shin, J.W.; Kim, H.; Cui, Y.; Kysar, J.W.; Hone, J.; Jung, Y.; Jeon, S.; et al. Strengthening effect of single-atomic-layer graphene in metal-graphene nanolayered composites. *Nat. Commun.* **2013**, *4*, 2114. [CrossRef] [PubMed]
14. Peng, Q.; Meng, F.; Yang, Y.; Lu, C.; Deng, H.; Wang, L.; De, S.; Gao, F. Shockwave generates < 100 > dislocation loops in bcc iron. *Nat. Commun.* **2018**, *9*, 4880. [PubMed]
15. Ferencz, R.; Sanchez, J.; Blumich, B.; Herrmann, W. AFM nanoindentation to determine Young's modulus for different EPDM elastomers. *Polym. Test* **2012**, *31*, 425–432. [CrossRef]
16. Liao, Q.Z.; Huang, J.Y.; Zhu, T.; Xiong, C.Y.; Fang, J. A hybrid model to determine mechanical properties of soft polymers by nanoindentation. *Mech. Mater.* **2010**, *42*, 1043–1047. [CrossRef]
17. Brenner, D.W.; Shenderova, O.A.; Harrison, J.A.; Stuart, S.J.; Ni, B.; Sinnott, S.B. A second-generation reactive empirical bond order (REBO) potential energy expression for hydrocarbons. *J. Phys.-Condens. Matter* **2002**, *14*, 783–802. [CrossRef]
18. Dewapriya, M.A.N.; Meguid, S.A. Atomistic modeling of out-of-plane deformation of a propagating Griffith crack in graphene. *Acta Mech.* **2017**, *228*, 3063–3075. [CrossRef]
19. Stuart, S.J.; Tutein, A.B.; Harrison, J.A. A reactive potential for hydrocarbons with intermolecular interactions. *J. Chem. Phys.* **2000**, *112*, 6472–6486. [CrossRef]
20. Daw, M.S.; Baskes, M.I. Embedded-Atom Method—Derivation and Application to Impurities, Surfaces, and Other Defects in Metals. *Phys. Rev. B* **1984**, *29*, 6443–6453. [CrossRef]
21. He, X.; Bai, Q.S.; Shen, R.Q. Atomistic perspective of how graphene protects metal substrate from surface damage in rough contacts. *Carbon* **2018**, *130*, 672–679. [CrossRef]
22. Yan, Y.; Zhou, S.; Liu, S. Atomistic simulation on nanomechanical response of indented graphene/nickel system. *Comput. Mater. Sci.* **2017**, *130*, 16–20. [CrossRef]

23. Hoover, W.G. Canonical Dynamics—Equilibrium Phase-Space Distributions. *Phys. Rev. A* **1985**, *31*, 1695–1697. [CrossRef]
24. Oliver, W.C.; Pharr, G.M. Measurement of hardness and elastic modulus by instrumented indentation: Advances in understanding and refinements to methodology. *J. Mater. Res.* **2004**, *19*, 3–20. [CrossRef]
25. Oliver, W.C.; Pharr, G.M. An Improved Technique for Determining Hardness and Elastic-Modulus Using Load and Displacement Sensing Indentation Experiments. *J. Mater. Res.* **1992**, *7*, 1564–1583. [CrossRef]
26. Marshall, D.B.; Noma, T.; Evans, A.G. A Simple Method for Determining Elastic-Modulus-to-Hardness Ratios Using Knoop Indentation Measurements. *J. Am. Ceram. Soc.* **1982**, *65*, C175–C176. [CrossRef]
27. Shang, N.G.; Papakonstantinou, P.; McMullan, M.; Chu, M.; Stamboulis, A.; Potenza, A.; Dhesi, S.S.; Marchetto, H. Catalyst-Free Efficient Growth, Orientation and Biosensing Properties of Multilayer Graphene Nanoflake Films with Sharp Edge Planes. *Adv. Funct. Mater.* **2010**, *18*, 3506–3514. [CrossRef]
28. Ghorbanzadeh Ahangari, M.; Fereidoon, A.; Hamed Mashhadzadeh, A. Interlayer interaction and mechanical properties in multi-layer graphene, Boron-Nitride, Aluminum-Nitride and Gallium-Nitride graphene-like structure: A quantum-mechanical DFT study. *Superlattices Microstruct.* **2017**, *112*, 30–45. [CrossRef]
29. Zhang, Y.Y.; Gu, Y.T. Mechanical properties of graphene: Effects of layer number, temperature and isotope. *Comp. Mater. Sci.* **2013**, *71*, 197–200. [CrossRef]
30. Shokrieh, M.M.; Hosseinkhani, M.R.; Naimi-Jamal, M.R.; Tourani, H. Nanoindentation and nanoscratch investigations on graphene-based nanocomposites. *Polym Test.* **2013**, *32*, 45–51. [CrossRef]
31. Mukhopadhyay, P.; Gupta, R.K. Trends and Frontiers in Graphene-Based Polymer Nanocomposites. *Plast. Eng.* **2011**, *67*, 32–42. [CrossRef]

© 2019 by the authors. Licensee MDPI, Basel, Switzerland. This article is an open access article distributed under the terms and conditions of the Creative Commons Attribution (CC BY) license (http://creativecommons.org/licenses/by/4.0/).

nanomaterials

MDPI

Article

Impact of Nano-Scale Distribution of Atoms on Electronic and Magnetic Properties of Phases in Fe-Al Nanocomposites: An Ab Initio Study

Ivana Miháliková [1,2], Martin Friák [1,*], Yvonna Jirásková [1], David Holec [3], Nikola Koutná [2,4] and Mojmír Šob [5,1,6]

1 Institute of Physics of Materials, Academy of Sciences of the Czech Republic, Žižkova 22, CZ-616 62 Brno, Czech Republic; ivanamihalik2@gmail.com (I.M.); jirasko@ipm.cz (Y.J.); mojmir@ipm.cz (M.Š.)
2 Department of Condensed Matter Physics, Faculty of Science, Masaryk University, Kotlářská 2, CZ-611 37 Brno, Czech Republic; nikakoutna@gmail.com
3 Department of Materials Science, Montanuniversität Leoben, Franz-Josef-Strasse 18, A-8700 Leoben, Austria; david.holec@unileoben.ac.at
4 Institute of Materials Science and Technology, TU Wien, Getreidemarkt 9, A-1060 Vienna, Austria
5 Department of Chemistry, Faculty of Science, Masaryk University, Kotlářská 2, CZ-611 37 Brno, Czech Republic
6 Central European Institute of Technology, CEITEC MU, Masaryk University, Kamenice 5, CZ-625 00 Brno, Czech Republic
* Correspondence: friak@ipm.cz; Tel.: +420-632-290-400

Received: 26 November 2018; Accepted: 13 December 2018; Published: 16 December 2018

Abstract: Quantum-mechanical calculations are applied to examine magnetic and electronic properties of phases appearing in binary Fe-Al-based nanocomposites. The calculations are carried out using the Vienna Ab-initio Simulation Package which implements density functional theory and generalized gradient approximation. The focus is on a disordered solid solution with 18.75 at. % Al in body-centered-cubic ferromagnetic iron, so-called α-phase, and an ordered intermetallic compound Fe_3Al with the $D0_3$ structure. In order to reveal the impact of the actual atomic distribution in the disordered Fe-Al α-phase three different special quasi-random structures with or without the 1st and/or 2nd nearest-neighbor Al-Al pairs are used. According to our calculations, energy decreases when eliminating the 1st and 2nd nearest neighbor Al-Al pairs. On the other hand, the local magnetic moments of the Fe atoms decrease with Al concentration in the 1st coordination sphere and increase if the concentration of Al atoms increases in the 2nd one. Furthermore, when simulating Fe-Al/Fe_3Al nanocomposites (superlattices), changes of local magnetic moments of the Fe atoms up to 0.5 μ_B are predicted. These changes very sensitively depend on both the distribution of atoms and the crystallographic orientation of the interfaces.

Keywords: Fe_3Al; Fe-Al; magnetism; interfaces; ab initio; stability; disorder

1. Introduction

Fe-Al-based materials represent one of the most promising classes of alloys intended for high-temperature applications. Remarkable are, in particular, their (i) resistance to oxidation, (ii) relatively low density, (iii) electrical resistivity and (iv) low cost of raw materials [1–3]. Their wider use is currently hindered by their lower ductility at ambient temperatures and a drop of the strength at elevated temperatures [3]. Regarding the former issue, the brittleness has been shown to be caused by an extrinsic effect, in particular hydrogen atoms [4,5] and there are experiments showing that the Fe_3Al could have reasonable ductility if the environmental embrittlement is eliminated [6,7].

In order to further fine-tune properties of iron-aluminium materials, they are intensively studied both experimentally and theoretically (see, for example, Refs. [8–10] or an excellent review of these activities published by Sundman and co-workers [11]). As far as basic thermodynamic properties are concerned, the Fe-Al phase diagram was originally determined by Kattner and Burton [12]. It reflects many phase transitions in the Fe-Al system from a disordered solid solution, A2 phase, through partially ordered, B2 phase, or ordered $D0_3$ phase. This offers some possibilities to produce two-phase Fe-Al alloy, e.g., ordered $D0_3$ phase embedded in B2 or disordered phase, the (nano)composite, exhibiting potentially new physical properties. From a viewpoint of applications, the final states of Fe-Al-based materials are highly sensitive to various factors including thermo-mechanical history (see, e.g., Ref. [13]). Again, it is because of many phase transitions in the Fe-Al system.

A particular sub-class of Fe-Al-based materials is represented by composites consisting of an ordered Fe_3Al phase with the $D0_3$ structure and a disordered Fe-Al solid solution with about 18–19 at.% Al. An experimental evidence of the co-existence of Fe_3Al and a disordered Fe-Al phase has been provided, for example, by transmission electron microscopy (TEM) technique which is sensitive to anti-phase boundaries (APBs). The APBs have a different character in Fe_3Al intermetalics and in the Fe-Al phase [14–17]. Specifically in the case of Fe-Al, the combined experimental study supported by theoretical simulations identified round/oval droplets of the disordered Fe-Al phase formed at the expense of diminishing amount of ordered Fe_3Al phase.

Next to the experimental research reported in the above mentioned papers, various Fe-Al-based materials were theoretically studied by quantum-mechanical calculations in the last three decades (see, e.g., Refs [18–33]). A surprising controversy has appeared in the case of first-principles prediction of the ground-state structure of Fe_3Al. For example, Lechermann et al. [34] determined the energies of Fe_3Al in the case of both experimentally observed $D0_3$ structure and a face-centered-cubic $L1_2$ one and showed that neither local density approximation (LDA) nor Perdew-Burke-Ernzerhof (PBE) [35] parametrization of the generalized gradient approximation (GGA) can correctly reproduce the $D0_3$ structure as the ground state of Fe_3Al. Similarly, Connetable and Maugis [36] calculated properties of Fe_3Al employing Perdew-Burke-Ernzerhof (PBE) parametrization [35] of the GGA and found out that Fe_3Al has lower energy in the $L1_2$ structure than in the experimentally observed $D0_3$ structure. In a subsequent paper by Lechermann and co-workers [37], electronic correlations and magnetism in Fe_3Al were studied employing LDA with an on-site Hubbard potential (LDA+U). The correct $D0_3$ structure was obtained as the ground state of Fe_3Al. As yet another example, Kellou et al. [38] compared the energies of Fe_3Al in both $D0_3$ and $L1_2$ structures using the pseudopotential plane-wave selfconsistent field ab-initio package and ultra-soft pseudopotentials. In this case, the correct ground-state structure was reproduced, too. Being aware of these findings, we have carefully chosen a reliable computational set up which provides a correct ground-state structure of Fe_3Al, i.e., the $D0_3$ structure.

Our paper aims at providing an analysis of complex relations between distributions of atoms on one hand and local magnetic moments of Fe atoms on the other in different Fe-Al phases. The complexity of the magnetic states is, in particular, exemplified by (i) the anti-correlations between the local magnetic moments of the Fe atoms and the number of Al atoms in the 1st coordination shell and (ii) a completely opposite trend, correlations between the local magnetic moments and the number of Al atoms in the 2nd coordination sphere of these Fe atoms. As the main novelty of our paper we show that the relations between local magnetic moments on one hand and the atomic distributions on the other hand are further complicated by the existence of interfaces in Fe-Al nanocomposites.

2. Methods

Our density-functional-theory [39,40] calculations employed projector augmented wave (PAW) pseudopotentials [41] and the exchange and correlation energy in the generalized gradient approximation (GGA) parametrized by Perdew and Wang [42] (PW91) with the Vosko-Wilk-Nusair correction [43] as implemented in the Vienna Ab initio Simulation Package (VASP) [44–46]. Plane-wave expansions were performed up to the cut-off energy of 350 eV and the Methfessel-Paxton method of the first order was adopted with a smearing width of 0.1 eV. The sampling of the Brillouin zone was done using Monkhorst-Pack [47] grids 10 × 10 × 6 and 10 × 10 × 3 for the simulation supercells containing 32 atoms ($2\sqrt{2} \times 2\sqrt{2} \times 2$ multiple of 2-atom cube-shape conventional bcc-cell) as models of individual phases (see Figure 1) and 64 atoms (double the size of 32-atom supercells—nanocomposites—see figures in Section 4). All local magnetic moments were initially oriented in a parallel manner which corresponds to the ferromagnetic state.

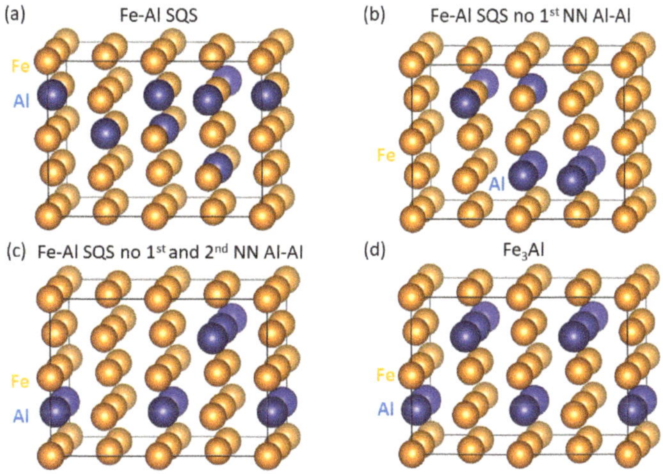

Figure 1. Schematic visualization of the 32-atom supercells used in our calculations: (**a**) a general special quasi-random structure (SQS) model for the Fe-Al phase (with 18.75 at.% Al), (**b**) an SQS model for the Fe-Al phase without any 1st nearest-neighbor (NN) Al-Al pairs, (**c**) an SQS model for the Fe-Al phase without the 1st and 2nd NN Al-Al pairs and (**d**) a 32-atom supercell of the Fe_3Al intermetallics. The unrelaxed atomic positions are listed in Table A1 in the Appendix.

3. Results for Individual Phases

When setting up computational supercells as models for phases appearing in Fe_3Al/Fe-Al nanocomposites, the disordered Fe-Al phase deserves a special attention. Our choice was motivated by our previous calculations of interactions of Al atoms in a bcc Fe ferromagnetic matrix [48] as well as by other theoretical results [49]. Regarding the latter, Amara and coworkers [49] studied the electronic structure and energetics of the dissolution of aluminum in α-iron and the interaction between Al atoms and vacancies. The stability of complexes containing Al and vacancy was found to be driven by strong Al-vacancy attractions and an Al-Al repulsion. Our calculations [48] also showed clear ordering tendencies of Al atoms. In particular, the energy of the system was decreasing with the elimination of the 1st and 2nd nearest neighbour (NN) Al-Al pairs, i.e., energies of the system containing the 1st and 2nd NN Al-Al pairs were higher than energies of the systems without them. Therefore, in this study, we compare properties obtained from three models which differ in the number of Al-Al pairs and we employed the concept of special quasi-random structure (SQS) [50] generated in USPEX code [51–53]. First, a general SQS (Figure 1a) containing the 1st and 2nd NN Al-Al pairs (A2-like with respect to Al-Al pairs) is used. Second, an SQS without any 1st NN Al-Al pairs (Figure 1b, B2-like w.r.t. Al-Al pairs) is utilized. Finally, an SQS without the 1st and the 2nd NN Al-Al pairs (effectively an Fe-rich Fe_3Al) (Figure 1c, i.e., $D0_3$-like w.r.t. the Al-Al pairs) is studied. We used 32-atom supercells which allow for a wider range of distribution of aluminium atoms in the disordered Fe-Al phase and the three different models have the stoichiometry $Fe_{26}Al_6$ (Figure 1a–c). The ordered intermetallic compound Fe_3Al is modeled by a 32-atom supercell with the stoichiometry $Fe_{24}Al_8$ (Figure 1d).

The thermodynamic stability of the Fe-Al polymorphs was assessed from their computed energies E by evaluating the formation energy: $E_f(Fe_xAl_y) = (E(Fe_xAl_y) - x \cdot E(Fe) - y \cdot E(Al))/(x+y)$ where x and y are numbers of Fe and Al atoms in the supercells and $E(Fe)$, $E(Al)$ are their chemical potentials, i.e., energies of elemental ferromagnetic (FM) body-centered cubic (bcc) Fe and non-magnetic (NM) face-centered cubic (fcc) Al. The computed formation energies, which partly appeared in Ref. [48], are −0.119 eV/atom (Figure 1a, A2-like), −0.121 eV/atom (Figure 1b, B2-like) and −0.144 eV/atom (Figure 1c, $D0_3$-like). The supercell with the least disordered distribution (Figure 1c) and Al atoms further apart is thus thermodynamically the most stable. This finding is in agreement with the Al-Al repulsion discussed in Refs. [48,49]. The other two (A2-like and B2-like) atomic distributions can be possibly considered as models for high-temperature states.

As one of the prime topics of our current study we examine relations between the value of local magnetic moments of Fe atoms and their surroundings. Figure 2 displays the spatial distribution of magnetic moments in the four studied phases with the diameter of spheres representing the magnitude of local magnetic moments. It is rather difficult to extract a clear pattern in the case of the three Fe-Al SQS polymorphs (Figure 2a–c) similarly as in our study of Fe-Al with different Al concentrations [20]. However, two different Fe sublattices in Fe_3Al with different values of local magnetic moment are easily recognizable in Figure 2d. One sublattice contains Fe atoms surrounded by 4 Al and 4 Fe atoms while the other has the Fe atoms surrounded by 8 Fe atoms. The local magnetic moments of the Fe atoms on the former sublattice are smaller (1.8 μ_B) than those corresponding to the latter case (2.4 μ_B).

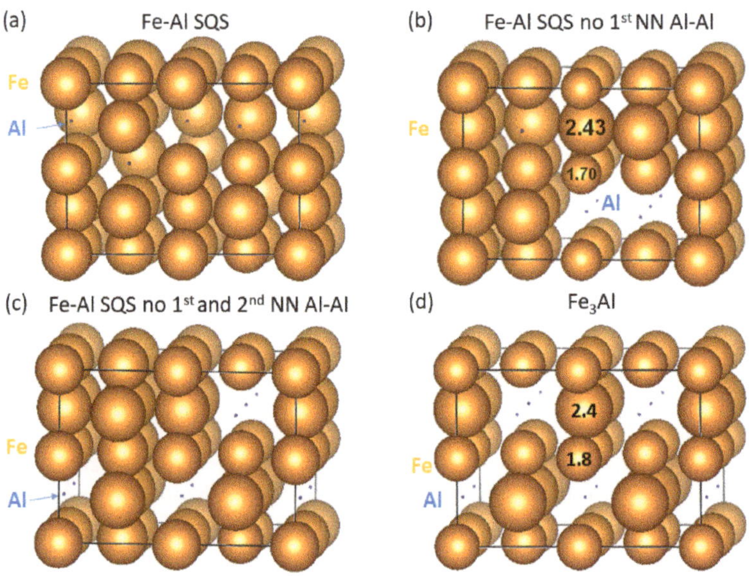

Figure 2. Schematic visualization of local atomic magnetic moments calculated for individual phases (see Figure 1): (**a–c**) are SQS models for the Fe-Al phase with different atomic arrangements and (**d**) for the Fe_3Al. The moments are visualized so that the diameters of the spheres reflect the magnitudes of the local magnetic moments (a few examples are in parts (**b**,**d**) in μ_B). Magnetic moments of Al atoms are so small, less than 0.05 μ_B, that they are shown only as blue dots.

In order to provide an overall description of all Fe atoms shown in Figure 1 and their corresponding local magnetic moments in Figure 2, we show them as functions of the concentration of Al atoms in Figure 3. In particular, the moments are found to decrease with increasing concentration of Al atoms in the 1st coordination shell, i.e., the fraction of 8 atoms in total in this shell, see Figure 3a.

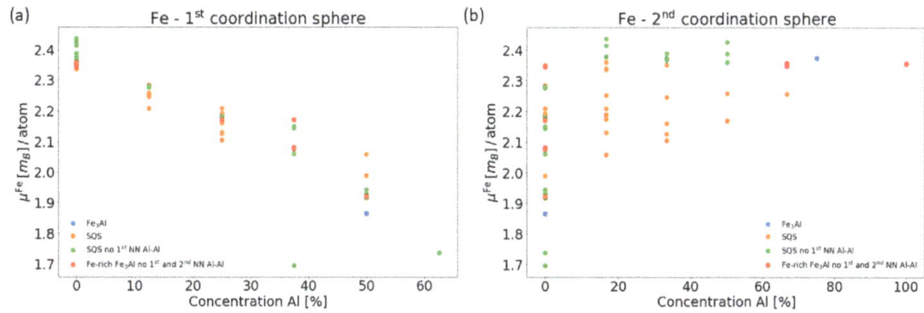

Figure 3. Dependences of local magnetic moments of Fe atoms in all studied phases (shown in Figure 1) as functions of concentration of Al atoms in the 1st (**a**) and the 2nd (**b**) coordination shell, respectively.

A similar finding, decreasing local magnetic moments with increasing number of Al atoms in the 1st nearest neighbor (NN) shell was also reported across a wider range of Al concentrations in Ref. [20]. Very interestingly, an opposite trend, i.e., an increase of the local magnetic moments of the Fe atoms as a function of the concentration of Al atoms (fraction out of 6 atoms in total), is obtained in the case of 2nd NN shell, see Figure 3b. These results clearly show a multi-faceted sensitivity of local magnetic moments of Fe atoms to the distribution of atoms in their local surroundings.

In order to shed light on the opposite trends of local magnetic moments of the Fe atoms as a function of the Al concentration in the 1st and 2nd coordination sphere, we recall the Stoner model which connects the value of the density of states at the Fermi level in a non-magnetic state with the tendency to spin polarization. We have treated all four studied phases (see Figure 1) as non-magnetic and determined the local DOS of individual Fe atoms at the Fermi level in the case of non-magnetic cases. The application of the Stoner model to individual atoms (see, e.g., a previous study [54]) is possible here as (i) contributions of Al atoms to the DOS at the Fermi level are small and (ii) the Stoner model is related to *d*-states rather than to *s*-/*p*-states. Our results are shown in Figure 4.

The values of DOS depicted in Figure 4 as functions of the Al concentration in the 1st or 2nd NN shell are similar to those shown in Figure 3. The local densities of states at the Fermi level of non-magnetic Fe atoms indeed resemble the trends of the local magnetic moments of these Fe atoms when they are spin-polarized, see Figure 4c.

As a next step we analyze local magnetic moments of Fe atoms as a function of their local atomic density of states at the Fermi level in the case of magnetic calculations (we consider the DOS as the sum of both spin channels). In contrast to the Stoner-like model elaborated above when we examined DOS of individual atoms in the non-magnetic state we below analyze the local DOSes of individual atoms in their spin-polarized, i.e., magnetic, states. The computed data points are shown in Figure 5. Both quantities are clearly anti-correlated. As they roughly follow linear trends, we use the least-square method to find suitable linear fitting functions. Interestingly, the slopes of these linear fitting functions, which are called parameter A in Figure 5, are significantly different for the three SQS Fe-Al variants (Figure 5a–c) on one hand and the ordered phase Fe_3Al (Figure 5d) on the other hand. There are differences in the slope among the three Fe-Al SQS variants, too, but differences within this group of three values ($A = -0.27$, -0.32 and -0.41 μ_B/state/eV) are relatively smaller than the difference between this group and the slope in the case of the Fe_3Al ($A = -0.58$). The difference between (i) the group of slope values related to the three Fe-Al SQS variants and (ii) the slope related to Fe_3Al can be tentatively attributed to the difference in the Al concentration (18.75 vs. 25 at.%). But the difference of the slopes of the three Fe-Al SQS variants is most likely a consequence of different distribution of atoms because the concentration of Al is equal in all three of them.

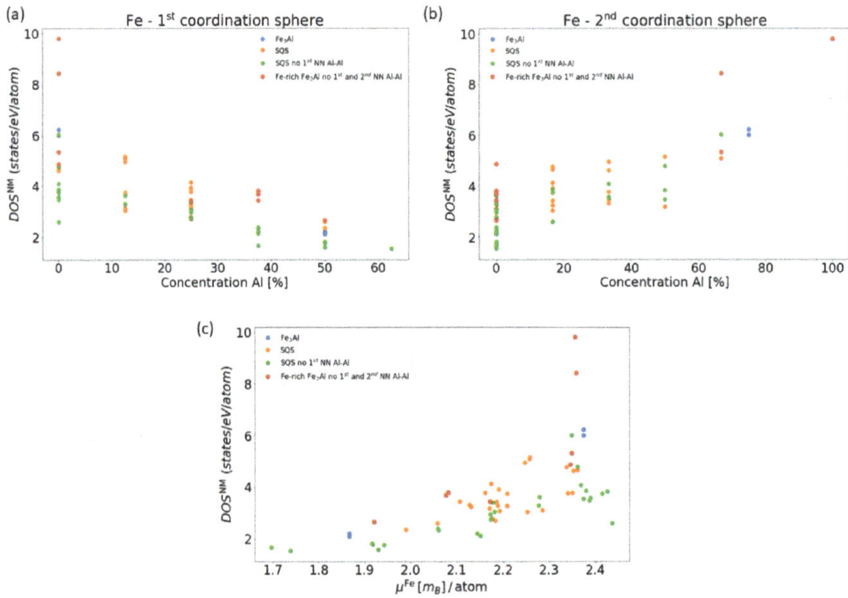

Figure 4. Calculated dependences of the density of states (DOS) of the Fe atoms at the Fermi level E_F in the case of NM states as a function of the Al concentration in the 1st (**a**) and the 2nd (**b**) coordination shell, respectively. Part (**c**) shows the DOS of the Fe atoms at the E_F in the case of NM states as a function of their local magnetic moments in the case of FM states.

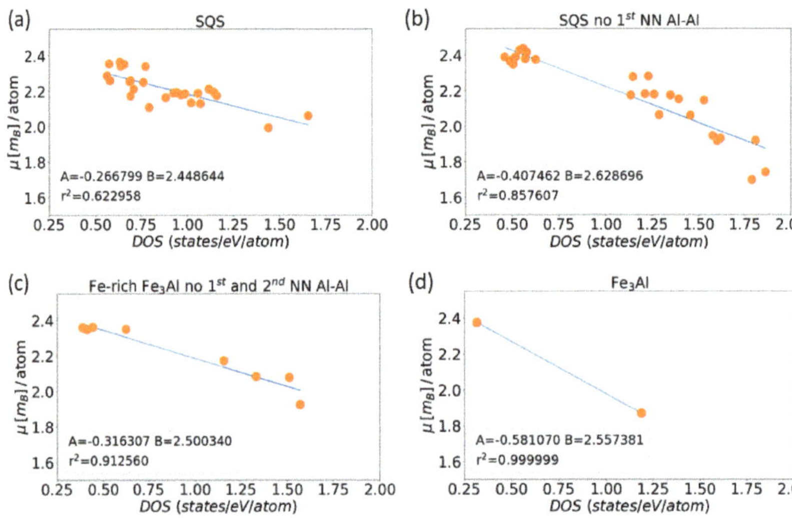

Figure 5. The dependences of local magnetic moments of Fe atoms on the density of states of these atoms (A and B represent coefficients in a linear fit $y = A \times x + B$ and r^2 is the coefficient of determination): (**a**) a general Fe-Al SQS, (**b**) an SQS without the first NN Al-Al pairs, (**c**) Fe-rich Fe_3Al without the first and the second nearest neighbour Al-Al pairs, and (**d**) Fe_3Al.

Besides, the local magnetic moments of Fe atoms differ significantly within the phases in question. This sensitivity of these moments is easily recognizable in the case of the ordered Fe_3Al phase where the moments are equal either to 1.8 or 2.4 μ_B. These different values stem from the existence of two different crystallographic sublattices of Fe atoms in the ordered Fe_3Al compound. Consequently, these sublattices represent qualitatively different chemical environments (4 Al + 4 Fe vs. 8 Fe atoms in the 1st coordination shell as discussed above). For the disordered Fe-Al SQS polymorphs the local magnetic moments cover wider ranges of values. For the general SQS the values of magnetic moments are between 1.9–2.4 μ_B, for the SQS without the 1st NN Al-Al pairs it is the widest obtained range from 1.6 to 2.45 μ_B and for the Fe-rich Fe_3Al without the 1st and the 2nd nearest neighbor Al-Al pairs the range is 1.8–2.4 μ_B.

The densities of states not only at the Fermi level but also at other energies are shown in Figure 6. The studied systems contain electrons with two opposite spin orientations which we further on refer to as UP and DOWN channels and analyze the DOS for each of them separately. In the case of spin UP-channel electrons of the four studied systems, the total DOSs (TDOSs) are qualitatively rather similar and do not exhibit any particularly noticeable features. In contrast, the spin DOWN-channel DOS exhibits specific trends. It appears that for the general SQS (Figure 6a) and the SQS without the 1st NN Al-Al pairs (Figure 6b) the densities of states have notable local maxima at the Fermi level when compared with that of the SQS without any 1st and 2nd NN Al-Al pairs (Figure 6c). This can be possibly linked to a higher thermodynamic stability (lower energy) of the SQS without the 1st and 2nd NN Al-Al pairs which was found in our previous study [48]. Note that as the DOWN-spin DOSs are depicted as negative values, the maxima appear as local minima.

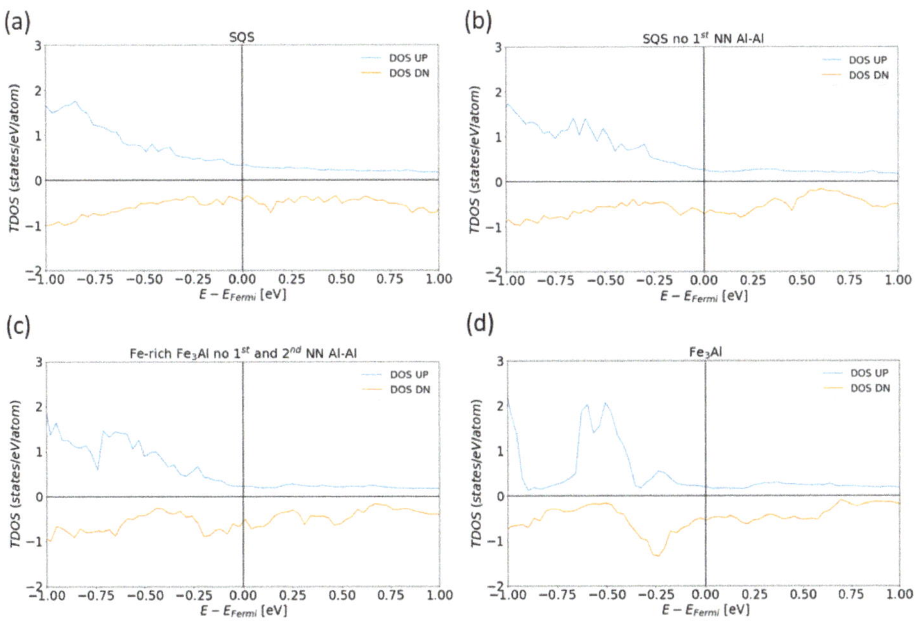

Figure 6. Calculated total densities of states (TDOS) in the case of the 32-atom supercells as models for individual phases: figures (**a**–**c**) are SQS models for the Fe-Al phase with different atomic arrangements and figure (**d**) for the Fe_3Al intermetallic compound.

After studying individual Fe-Al and Fe_3Al phases separately, we next examine changes induced in the local magnetic moments of the Fe atoms when forming their nanocomposites.

4. Results for Nanocomposites

The supercells of individual phases shown in Figure 1 have the facets with (001), (110) and ($1\bar{1}0$) crystallographic orientations (with respect to a two-atom conventional cubic cell of bcc Fe). Therefore, when combining three different polymorphs of Fe-Al with the Fe_3Al intermetallic compound three different nanocomposites with different interfaces between the phases can be simulated. They are schematically visualized in Figures 7 and 8 (the construction in the case of the ($1\bar{1}0$) interface plane is not shown as it is rather similar to that of the (110) one). It is worth noting that due to the periodic boundary conditions applied in our calculations, the simulated nanocomposites form so-called superlattices [55–76] when both phases coherently co-exist and the atomic planes continue from one phase into another. As another consequence of (i) the periodicity and (ii) the fact that the supercells modeling the Fe-Al phase are disordered, the two interfaces per 64-atom supercell are not the same. The scalar properties (e.g., the interface energy γ below) are then averages of the two interfaces.

Regarding the thermodynamic properties of the studied interfaces, they were assessed in Ref. [48]. In particular, the interface energies γ (in fact, their averages—see the discussion above) were evaluated according to the formula $\gamma = (E(Fe_3Al/Fe\text{-}Al) - E(Fe_3Al) - E(Fe\text{-}Al))/(2S)$ from the energy of the composite system $E(Fe_3Al/Fe\text{-}Al)$, the energies of individual phases, $E(Fe_3Al)$ and $E(Fe\text{-}Al)$ and the area of the interfaces S. The interface energies turned out to be very weakly dependent on the crystallographic orientation of interfaces (see Ref. [48]). For example, for the nanocomposites containing the general SQS, see Figure 1a, the interface energies are equal to 0.019 J/m^2, 0.020 J/m^2 and 0.022 J/m^2 for the (001), (110) and ($1\bar{1}0$) orientation, respectively. It should be noted that these very low interface energies are quite close to an estimated error-bar of our calculations, about 0.005 J/m^2.

The fact that the interface energies are so similar for different orientations is in line with the findings of Oguma et al. [17] who identified round/oval droplets of the disordered Fe-Al phase surrounded by the Fe_3Al phase. The rounded shape of these droplets can be probably connected with the fact that interface energy is not sensitive to crystallographic orientation and, therefore, the interfaces studied in this paper are equally probable as others. It should be noted that other properties, such as mechanical ones, can be much more sensitive with respect to the orientation. Further the interface energies sensitively depend on the distribution of atoms and become practically zero (within the error-bar of our calculations, i.e., about 0.001 J/m^2) for nanocomposites containing Fe_3Al and the Fe-Al variant without the 1st and 2nd NN Al-Al pairs (a Fe-rich variant of the Fe_3Al). This unusual result can be explained by the fact that Fe_3Al can contain rather high concentration of point defects (such as off-stoichiometric Al atom, anti-sites) and covers rather broad range of Al compositions in the Fe-Al phase diagram around 25 at.%. The simulated nanocomposites are then quite similar to a single-phase material (Fe_3Al with point defects) containing perfect and defected regions.

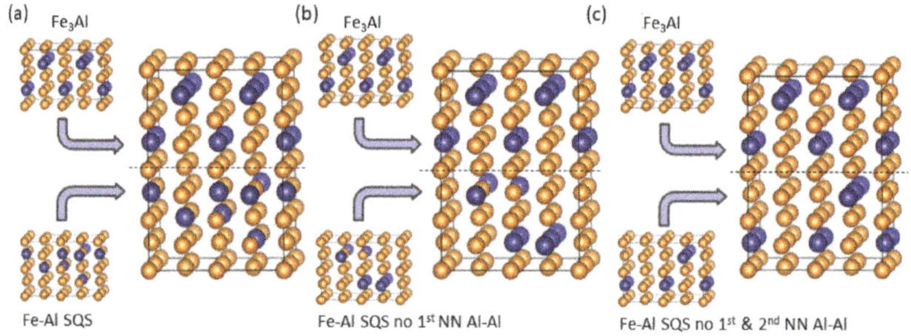

Figure 7. Visualization of nanocomposite supercells for the (001) interface plane. The figures show how the Fe_3Al intermetallic compound is combined with different models of the Fe-Al phase with 18.75 at.% Al, in particular the general SQS (A2-like) model (**a**), the SQS model without the 1st NN Al-Al pairs, B2-like, (**b**) and the SQS without the 1st and 2nd NN Al-Al pairs, $D0_3$-like (**c**).

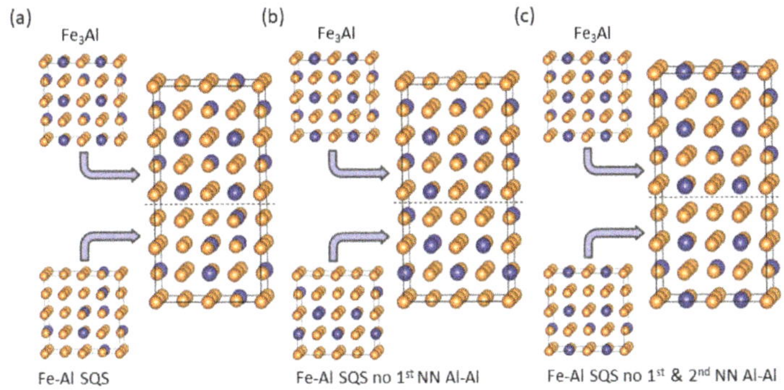

Figure 8. The same as in Figure 7 but for nanocomposite supercells with the (110) interface plane.

Figure 9 shows the differences in the magnitude of local magnetic moments of Fe atoms induced by the interfaces for all three variants of the Fe-Al phase and the three orientations of interfaces.

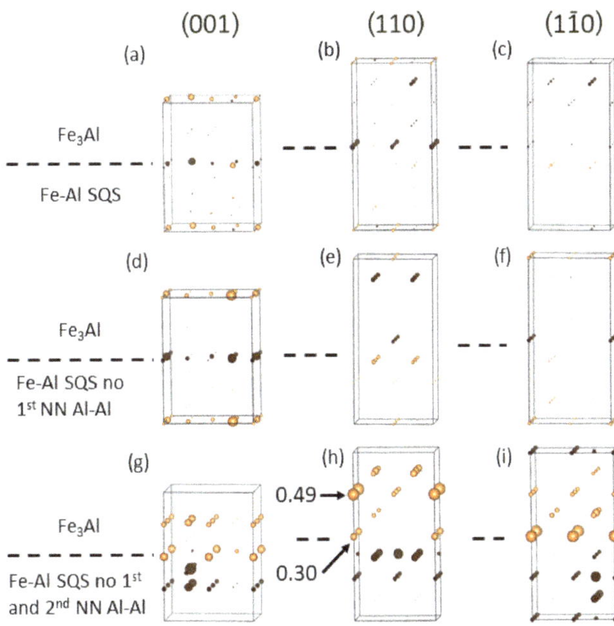

Figure 9. Differences in the magnitude of local magnetic moments of Fe atoms induced by the interfaces within the nanocomposites. They are visualized by the diameter of spheres representing individual Fe atoms (two are listed in subfigure (**h**)). Gold (black) color of the spheres indicate positive (negative) changes. The sub-figures (**a**,**d**,**g**) correspond to the nanocomposites shown in Figure 7a–c, respectively, the sub-figures (**b**,**e**,**h**) to those in Figure 8a–c, respectively, and the sub-figures (**c**,**f**,**i**) corresponds to the nanocomposites with the ($1\bar{1}0$) orientation of the interfaces (not shown). Please note that due to the fact that the differences are rather small, the scaling connecting the value of the difference and the diameter of the spheres is three times bigger than the scaling applied in Figure 2.

The changes are indicated by the diameter of spheres representing individual atoms. They are rather small, up to about 0.5 μ_B. Let us note that in order to make the spheres representing individual Fe atoms in Figure 9 visible also in the case of very small values of the changes, the scaling connecting (i) the value of the difference on one hand and (ii) the diameter of the sphere in Figure 9 on the other hand, is three times bigger than the scaling applied in Figure 2 above. When inspecting the changes for different orientation of interfaces (rows of subfigures in Figure 9) we find that they depend on the orientation and when analyzing them for different models of the Fe-Al phase (columns of subfigures in Figure 9) we identify that the magnetic moments of Fe atoms sensitively respond to the differences in the distribution of atoms. But the actual changes are clearly rather small.

5. Conclusions

We have performed a series of ab initio calculations to examine magnetic and electronic properties of two phases appearing in binary Fe-Al-based nanocomposites. In particular, a disordered solid solution with 18.75 at. % Al in body-centered-cubic (bcc) ferromagnetic (FM) iron, so-called α-phase, was studied together with the ordered intermetallic compound Fe_3Al. By comparing results for three different special quasi-random structures (SQS) for the α-phase which differ in the distribution of atoms (they are with or without 1st and/or 2nd nearest-neighbor Al-Al pairs) we found the local magnetic moments of iron atoms clearly affected by the chemical composition of neighboring coordination shells. In particular, the local magnetic moments decrease (increase) with the concentration of Al in the 1st (2nd) coordination shell.

In connection with the Stoner model, a similar tendencies were found in the density of states of individual Fe atoms at the Fermi level as a function of the Al concentration in the 1st and 2nd NN shell. Further, when simulating Fe-Al/Fe_3Al nanocomposites (superlattices) changes of local magnetic moments of the Fe atoms (up to 0.5 μ_B) are found but they depend sensitively on both the distribution of atoms in the Fe-Al α-phase and the crystallographic orientation of the interfaces. Our findings aim at stimulating further research of coherent nanocomposites of magnetic materials, for example experimental studies employing the Extended X-Ray Absorption Fine Structure (EXAFS) technique and, in particular, the multiple-scattering approach to EXAFS analysis, called GNXAS [77,78], with which it is possible to calculate two-, three- and four-atom correlation functions.

Author Contributions: Conceptualization, I.M. and M.F.; Methodology, I.M, D.H. and N.K.; Resources, M.Š.; Writing—Original Draft Preparation, I.M. and M.F.; Writing—Review & Editing, I.M.,N.K., D.H., Y.J., and M.Š.; Visualization, M.F. and I.M.; Supervision, M.Š., M.F., Y.J.; Project Administration and Funding Acquisition, M.Š and M.F.

Funding: The authors acknowledge the Czech Science Foundation for the financial support received under the Projects Nos. 16-24711S (M.Š. and N.K.) and 17-22139S (M.F., Y.J. and I.M.). Additional resources were provided by the Academy of Sciences of the Czech Republic through the Fellowship of J. E. Purkyně (M.F.) and by the Ministry of Education, Youth and Sports of the Czech Republic under the Project CEITEC 2020, LQ1601 (M.Š).

Acknowledgments: M.F., I.M., and M.Š. also acknowledge supports from the Academy of Sciences of the Czech Republic (Institutional Project No. RVO:68081723) and the Ministry of Education, Youth and Sports of the Czech Republic via the research infrastructure IPMINFRA, LM2015069. Computational resources were made available by the Ministry of Education, Youth and Sports of the Czech Republic under the Projects CESNET (Project No. LM2015042), CERIT-Scientific Cloud (Project No. LM2015085) and IT4Innovations National Supercomputer Center (Project No. LM2015070) within the program Projects of Large Research, Development and Innovations Infrastructures. Special quasi-random distributions of atoms were obtained using the USPEX code [51–53], based on an evolutionary algorithm developed by Oganov, Glass, Lyakhov and Zhu and featuring local optimization, real-space representation and flexible physically motivated variation operators. Figures 1, 2, 7–9 were visualized employing the VESTA package [79].

Conflicts of Interest: The authors declare no conflict of interest.

Appendix A

Table A1. Atomic positions (expressed as fractions of supercell dimensions) within the computational supercells shown in Figure 1. Aluminium positions are the first six rows in the case of Fe-Al variants and the first eight rows in the case of Fe_3Al.

Fe-Al SQS			Fe-Al SQS no 1NN			Fe-Al SQS no 1&2NN			Fe_3Al		
1	1	$\sqrt{2}/2$	1	1	$\sqrt{2}/2$	1	1	$\sqrt{2}/2$	1	1	$\sqrt{2}/2$
0.75	0.75	0.50	0.50	0.75	0.25	0.25	0.00	0.75	0.25	0.00	0.75
0.25	0.00	0.75	0.75	0.00	0.75	0.25	0.50	0.75	0.75	0.00	0.75
0.50	0.75	0.75	0.25	0.00	0.25	0.00	0.25	0.25	0.25	0.50	0.75
0.00	0.75	0.75	0.25	0.50	0.25	0.00	0.75	0.25	0.75	0.50	0.75
0.25	0.50	0.25	0.50	0.25	0.25	0.50	0.25	0.25	0.00	0.25	0.25
0.50	0.50	0.50	0.50	0.25	0.75	0.50	0.75	0.25	0.00	0.75	0.25
0.50	0.25	0.25	0.75	0.50	0.75	0.75	0.50	0.75	0.50	0.25	0.25
0.75	0.50	0.25	0.00	0.75	0.75	0.75	0.00	0.75	0.50	0.75	0.25
0.00	0.25	0.75	0.50	0.75	0.75	0.00	0.25	0.75	0.00	0.25	0.75
0.00	0.75	0.25	0.00	0.75	0.25	0.00	0.75	0.75	0.00	0.75	0.75
0.00	0.00	0.00	0.00	0.25	0.25	0.50	0.25	0.75	0.50	0.25	0.75
0.00	0.25	0.25	0.75	0.00	0.25	0.50	0.75	0.75	0.50	0.75	0.75
0.25	0.00	0.25	0.25	0.50	0.75	0.25	0.00	0.25	0.25	0.00	0.25
0.50	0.75	0.25	0.75	0.50	0.25	0.75	0.00	0.25	0.75	0.00	0.25
0.75	0.75	0.00	0.25	0.00	0.75	0.25	0.50	0.25	0.25	0.50	0.25
0.25	0.50	0.75	0.00	0.25	0.75	0.75	0.50	0.25	0.75	0.50	0.25
0.75	0.50	0.75	0.75	0.75	0.50	0.00	0.00	0.00	0.00	0.00	0.00
0.00	0.50	0.00	0.50	0.50	0.50	0.50	0.00	0.00	0.50	0.00	0.00
0.50	0.25	0.75	0.00	0.00	0.00	0.00	0.50	0.00	0.00	0.50	0.00
0.75	0.00	0.25	0.75	0.75	0.00	0.50	0.50	0.00	0.50	0.50	0.00
0.25	0.25	0.50	0.00	0.50	0.00	0.00	0.00	0.50	0.00	0.00	0.50
0.50	0.00	0.50	0.25	0.25	0.50	0.50	0.00	0.50	0.50	0.00	0.50
0.00	0.00	0.50	0.50	0.00	0.50	0.00	0.50	0.50	0.00	0.50	0.50
0.50	0.50	0.00	0.00	0.00	0.50	0.50	0.50	0.50	0.50	0.50	0.50
0.25	0.25	0.00	0.50	0.50	0.00	0.25	0.25	0.00	0.25	0.25	0.00
0.75	0.25	0.00	0.25	0.25	0.00	0.75	0.25	0.00	0.75	0.25	0.00
0.25	0.75	0.00	0.75	0.25	0.00	0.25	0.75	0.00	0.25	0.75	0.00
0.75	0.00	0.75	0.25	0.75	0.00	0.75	0.75	0.00	0.75	0.75	0.00
0.50	0.00	0.00	0.50	0.00	0.00	0.25	0.25	0.50	0.25	0.25	0.50
0.75	0.25	0.50	0.75	0.25	0.50	0.75	0.25	0.50	0.75	0.25	0.50
0.25	0.75	0.50	0.25	0.75	0.50	0.25	0.75	0.50	0.25	0.75	0.50
0.00	0.50	0.50	0.00	0.50	0.50	0.75	0.75	0.50	0.75	0.75	0.50

References

1. Sauthoff, G. *Intermetallics*; VCH Verlagsgesellschaft: Weinheim, Germany, 1995.
2. Liu, C.T.; Stringer, J.; Mundy, J.N.; Horton, L.L.; Angelini, P. Ordered intermetallic alloys: An assessment. *Intermetallics* **1997**, *5*, 579–596. [CrossRef]
3. Stoloff, N.S. Iron aluminides: Present status and future prospects. *Mater. Sci. Eng. A* **1998**, *258*, 1–14. [CrossRef]
4. Liu, C.T.; Lee, E.H.; McKamey, C.G. An environmental-effect as the major cause for room-temperature embrittlement in FeAl. *Scr. Metall.* **1989**, *23*, 875–880. [CrossRef]
5. Lynch, R.J.; Heldt, L.A.; Milligan, W.W. Effects of alloy composition on environmental embrittlement of B2 ordered iron aluminides. *Scr. Metall.* **1991**, *25*, 2147–2151. [CrossRef]
6. Liu, C.T.; McKamey, C.G.; Lee, E.H. Environmental-effects on room-temperature ductility and fracture in Fe_3Al. *Scr. Metall.* **1990**, *24*, 385–389. [CrossRef]
7. Lynch, R.J.; Gee, K.A.; Heldt, L.A. Environmental embrittlement of single-crystal and thermomechanically processed B2-ordered iron aluminides. *Scr. Metall.* **1994**, *30*, 945–950. [CrossRef]
8. Palm, M.; Inden, G.; Thomas, N. The Fe-Al-Ti system. *J. Phase Equilib.* **1995**, *16*, 209–222. [CrossRef]

9. Palm, M.; Lacaze, J. Assessment of the Al-Fe-Ti system. *Intermetallics* **2006**, *14*, 1291–1303. [CrossRef]
10. Palm, M.; Sauthoff, G. Deformation behaviour and oxidation resistance of single-phase and two-phase $L2_1$-ordered Fe-Al-Ti alloys. *Intermetallics* **2004**, *12*, 1345–1359. [CrossRef]
11. Sundman, B.; Ohnuma, I.; Dupin, N.; Kattner, U.R.; Fries, S.G. An assessment of the entire Al-Fe system including D0(3) ordering. *Acta Mater.* **2009**, *57*, 2896–2908. [CrossRef]
12. Kattner, U.; Burton, B. Al-Fe (Aluminium-Iron). In *Phase Diagrams of Binary Iron Alloys*; Okamoto, H., Ed.; ASM International: Geauga County, OH, USA, 1993, pp. 12–28.
13. Allen, S.; Cahn, J. Mechanisms of phase-transformations within miscibility gap of Fe-rich Fe-Al alloys. *Acta Metall. Mater.* **1976**, *24*, 425–437. [CrossRef]
14. Wang, K.; Wang, Y.; Cheng, Y. The Formation and Dynamic Evolution of Antiphase Domain Boundary in FeAl Alloy: Computational Simulation in Atomic Scale. *Mater. Res.-Ibero-Am. J. Mater.* **2018**, *21*. [CrossRef]
15. Balagurov, A.M.; Bobrikov, I.A.; Sumnikov, V.S.; Golovin, I.S. Antiphase domains or dispersed clusters? Neutron diffraction study of coherent atomic ordering in Fe_3Al-type alloys. *Acta Mater.* **2018**, *153*, 45–52. [CrossRef]
16. Murakami, Y.; Niitsu, K.; Tanigaki, T.; Kainuma, R.; Park, H.S.; Shindo, D. Magnetization amplified by structural disorder within nanometre-scale interface region. *Nat. Commun.* **2014**, *5*. [CrossRef] [PubMed]
17. Oguma, R.; Matsumura, S.; Eguchi, T. Kinetics of B2-and $D0_3$ type ordering and formation of domain structures in Fe-Al alloys. *J. Phys. Condens. Matter* **2008**, *20*, 275225. [CrossRef] [PubMed]
18. Watson, R.E.; Weinert, M. Transition-metal aluminide formation: Ti, V, Fe, and Ni aluminides. *Phys. Rev. B* **1998**, *58*, 5981–5988. [CrossRef]
19. Gonzales-Ormeno, P.; Petrilli, H.; Schon, C. Ab-initio calculations of the formation energies of BCC-based superlattices in the Fe-Al system. *Calphad* **2002**, *26*, 573. [CrossRef]
20. Friák, M.; Neugebauer, J. Ab initio study of the anomalous volume-composition dependence in Fe-Al alloys. *Intermetallics* **2010**, *18*, 1316–1321. [CrossRef]
21. Friák, M.; Oweisová, S.; Pavlů, J.; Holec, D.; Šob, M. An Ab Initio Study of Thermodynamic and Mechanical Stability of Heusler-Based Fe_2AlCo Polymorphs. *Materials* **2018**, *11*, 1543. [CrossRef]
22. Friák, M.; Holec, D.; Šob, M. Quantum-Mechanical Study of Nanocomposites with Low and Ultra-Low Interface Energies. *Nanomaterials* **2018**, *8*, 1057. [CrossRef]
23. Friák, M.; Slávik, A.; Miháliková, I.; Holec, D.; Všianská, M.; Šob, M.; Palm, M.; Neugebauer, J. Origin of the Low Magnetic Moment in Fe_2AlTi: An Ab Initio Study. *Materials* **2018**, *11*, 1732. [CrossRef]
24. Šesták, P.; Friák, M.; Holec, D.; Všianská, M.; Šob, M. Strength and Brittleness of Interfaces in Fe-Al Superalloy Nanocomposites under Multiaxial Loading: An ab initio and Atomistic Study. *Nanomaterials* **2018**, *8*, 873. [CrossRef] [PubMed]
25. Liu, S.; Duan, S.; Ma, B. First-principles calculation of vibrational entropy for Fe-Al compounds. *Phys. Rev. B* **1998**, *58*, 9705–9709.
26. Kulikov, N.I.; Postnikov, A.V.; Borstel, G.; Braun, J. Onset of magnetism in B2 transition-metal aluminides. *Phys. Rev. B* **1999**, *59*, 6824–6833. [CrossRef]
27. Fähnle, M.; Drautz, R.; Lechermann, F.; Singer, R.; Diaz-Ortiz, A.; Dosch, H. Thermodynamic properties from ab-initio calculations: New theoretical developments, and applications to various materials systems. *Phys. Status Solidi B Basic Solid State Phys.* **2005**, *242*, 1159–1173. [CrossRef]
28. Friák, M.; Deges, J.; Krein, R.; Frommeyer, G.; Neugebauer, J. Combined ab initio and experimental study of structural and elastic properties of Fe_3Al-based ternaries. *Intermetallics* **2010**, *18*, 1310. [CrossRef]
29. Kirklin, S.; Saal, J.E.; Hegde, V.I.; Wolverton, C. High-throughput computational search for strengthening precipitates in alloys. *Acta Mater.* **2016**, *102*, 125–135. [CrossRef]
30. Airiskallio, E.; Nurmi, E.; Heinonen, M.H.; Vayrynen, I.J.; Kokko, K.; Ropo, M.; Punkkinen, M.P.J.; Pitkanen, H.; Alatalo, M.; Kollar, J.; et al. High temperature oxidation of Fe-Al and Fe-Cr-Al alloys: The role of Cr as a chemically active element. *Corros. Sci.* **2010**, *52*, 3394 – 3404. [CrossRef]
31. Medvedeva, N.I.; Park, M.S.; Van Aken, D.C.; Medvedeva, J.E. First-principles study of Mn, Al and C distribution and their effect on stacking fault energies in fcc Fe. *J. Alloy. Compd.* **2014**, *582*, 475–482. [CrossRef]
32. Čížek, J.; Lukáč, F.; Procházka, I.; Kužel, R.; Jirásková, Y.; Janičkovič, D.; Anwand, W.; Brauer, G. Characterization of quenched-in vacancies in Fe-Al alloys. *Phys. B* **2012**, *407*, 2659–2664. [CrossRef]

33. Ipser, H.; Semenova, O.; Krachler, R. Intermetallic phases with D0(3)-structure: A statistical-thermodynamic model. *J. Alloy. Compd.* **2002**, *338*, 20–25. [CrossRef]
34. Lechermann, F.; Welsch, F.; Elsässer, C.; Ederer, C.; Fähnle, M.; Sanchez, J.; Meyer, B. Density-functional study of Fe_3Al: LSDA versus GGA. *Phys. Rev. B* **2002**, *65*, 132104. [CrossRef]
35. Perdew, J.P.; Burke, K.; Ernzerhof, M. Generalized Gradient Approximation Made Simple. *Phys. Rev. Lett.* **1996**, *77*, 3865–3868. [CrossRef] [PubMed]
36. Connetable, D.; Maugis, P. First principle calculations of the kappa-Fe_3AlC perovskite and iron-aluminium intermetallics. *Intermetallics* **2008**, *16*, 345–352. [CrossRef]
37. Lechermann, F.; Fähnle, M.; Meyer, B.; Elsässer, C. Electronic correlations, magnetism, and structure of Fe-Al subsystems: An LDA+U study. *Phys. Rev. B* **2004**, *69*, 165116. [CrossRef]
38. Kellou, A.; Grosdidier, T.; Raulot, J.M.; Aourag, H. Atomistic study of magnetism effect on structural stability in Fe_3Al and Fe_3AlX (X = H, B, C, N, O) alloys. *Phys. Status Solidi B Basic Solid State Phys.* **2008**, *245*, 750–755. [CrossRef]
39. Hohenberg, P.; Kohn, W. Inhomogeneous electron gas. *Phys. Rev. B* **1964**, *136*, B864–B871. [CrossRef]
40. Kohn, W.; Sham, L.J. Self-consistent equations including exchange and correlation effects. *Phys. Rev. A* **1965**, *140*, A1133–A1138. [CrossRef]
41. Blöchl, P.E. Projector augmented-wave method. *Phys. Rev. B* **1994**, *50*, 17953–17979. [CrossRef]
42. Perdew, J.P.; Wang, Y. Accurate and simple analytic representation of the electron-gas correlation energy. *Phys. Rev. B* **1992**, *45*, 13244–13249. [CrossRef]
43. Vosko, S.H.; Wilk, L.; Nusair, M. Accurate spin-dependent electron liquid correlation energies for local spin density calculations: A critical analysis. *Can. J. Phys.* **1980**, *58*, 1200. [CrossRef]
44. Kresse, G.; Hafner, J. Ab initio molecular dynamics for liquid metals. *Phys. Rev. B* **1993**, *47*, 558–561. [CrossRef]
45. Kresse, G.; Furthmüller, J. Efficient iterative schemes for ab initio total-energy calculations using a plane-wave basis set. *Phys. Rev. B* **1996**, *54*, 11169–11186. [CrossRef]
46. Kresse, G.; Joubert, D. From ultrasoft pseudopotentials to the projector augmented-wave method. *Phys. Rev. B* **1999**, *59*, 1758–1775. [CrossRef]
47. Monkhorst, H.J.; Pack, J.D. Special points for Brillouin-zone integrations. *Phys. Rev. B* **1976**, *13*, 5188–5192. [CrossRef]
48. Mihálikova, I.; Slávik, A.; Friák, M.; Všianská, M.; Koutná, N.; Holec, D.; Šob, M. First-principles study of interface energies in Fe-Al-based superalloy nanocomposites. In Proceedings of the NANOCON 9th International Conference on Nanomaterials—Research & Application, Brno, Czech Republic, 18–20 October 2017; Tanger Ltd.: Ostrava, Czech Republic 2018; pp. 69–74.
49. Amara, H.; Fu, C.C.; Soisson, F.; Maugis, P. Aluminum and vacancies in α-iron: Dissolution, diffusion, and clustering. *Phys. Rev. B* **2010**, *81*, 174101. [CrossRef]
50. Zunger, A.; Wei, S.; Ferreira, L.; Bernard, J. Special quasirandom structures. *Phys. Rev. Lett.* **1990**, *65*, 353–356. [CrossRef]
51. Oganov, A.R.; Glass, C.W. Crystal structure prediction using ab initio evolutionary techniques: Principles and applications. *J. Chem. Phys.* **2006**, *124*, 244704. [CrossRef]
52. Lyakhov, A.O.; Oganov, A.R.; Stokes, H.T.; Zhu, Q. New developments in evolutionary structure prediction algorithm USPEX. *Comput. Phys. Commun.* **2013**, *184*, 1172–1182. [CrossRef]
53. Oganov, A.R.; Lyakhov, A.O.; Valle, M. How Evolutionary Crystal Structure Prediction Works—And Why. *Acc. Chem. Res.* **2011**, *44*, 227–237. [CrossRef]
54. Bose, S.K.; Kudrnovský, J.; Drchal, V.; Turek, I. Magnetism of mixed quaternary Heusler alloys: $(\mathrm{Ni}, T)_2\mathrm{MnSn}$ $(T = \mathrm{Cu}, \mathrm{Pd})$ as a case study. *Phys. Rev. B* **2010**, *82*, 174402. [CrossRef]
55. Mayrhofer, P.H.; Fischer, F.D.; Boehm, H.J.; Mitterer, C.; Schneider, J.M. Energetic balance and kinetics for the decomposition of supersaturated Ti1-xAlxN. *Acta Mater.* **2007**, *55*, 1441–1446. [CrossRef]
56. Wu, L.; Chen, M.; Li, C.; Zhou, J.; Shen, L.; Wang, Y.; Zhong, Z.; Feng, M.; Zhang, Y.; Han, K.; et al. Ferromagnetism and matrix-dependent charge transfer in strained $LaMnO_3$-$LaCoO_3$ superlattices. *Mater. Res. Lett.* **2018**, *6*, 501–507. [CrossRef]

57. Koutná, N.; Holec, D.; Friák, M.; Mayrhofer, P.H.; Šob, M. Stability and elasticity of metastable solid solutions and superlattices in the MoN–TaN system: First-principles calculations. *Mater. Des.* **2018**, *144*, 310–322. [CrossRef]
58. Jiang, M.; Xiao, H.Y.; Peng, S.M.; Yang, G.X.; Liu, Z.J.; Zu, X.T. A comparative study of low energy radiation response of AlAs, GaAs and GaAs/AlAs superlattice and the damage effects on their electronic structures. *Sci. Rep.* **2018**, *8*, 2012. [CrossRef]
59. Wen, Y.N.; Gao, P.F.; Xia, M.G.; Zhang, S.L. Half-metallic ferromagnetism prediction in MoS_2-based two-dimensional superlattice from first-principles. *Mod. Phys. Lett. B* **2018**, *32*, 1850098. [CrossRef]
60. Friák, M.; Tytko, D.; Holec, D.; Choi, P.P.; Eisenlohr, P.; Raabe, D.; Neugebauer, J. Synergy of atom-probe structural data and quantum-mechanical calculations in a theory-guided design of extreme-stiffness superlattices containing metastable phases. *New J. Phys.* **2015**, *17*, 093004. [CrossRef]
61. Dai, Q.; Eckern, U.; Schwingenschlog, U. Effects of oxygen vacancies on the electronic structure of the (LaVO3)(6)/SrVO3 superlattice: A computational study. *New J. Phys.* **2018**, *20*, 073011. [CrossRef]
62. Jiang, M.; Xiao, H.; Peng, S.; Qiao, L.; Yang, G.; Liu, Z.; Zu, X. First-Principles Study of Point Defects in GaAs/AlAs Superlattice: The Phase Stability and the Effects on the Band Structure and Carrier Mobility. *Nanoscale Res. Lett.* **2018**, *13*, 301. [CrossRef]
63. Chen, H.; Millis, A.J.; Marianetti, C.A. Engineering Correlation Effects via Artificially Designed Oxide Superlattices. *Phys. Rev. Lett.* **2013**, *111*, 116403. [CrossRef]
64. Mottura, A.; Janotti, A.; Pollock, T.M. A first-principles study of the effect of Ta on the superlattice intrinsic stacking fault energy of $L1_2$-Co-3(Al,W). *Intermetallics* **2012**, *28*, 138–143. [CrossRef]
65. Rosengaard, N.; Skriver, H. Ab-initio study of antiphase boundaries and stacking-faults in $L1_2$ and $D0_{22}$ compounds. *Phys. Rev. B* **1994**, *50*, 4848–4858. [CrossRef]
66. Torres-Pardo, A.; Gloter, A.; Zubko, P.; Jecklin, N.; Lichtensteiger, C.; Colliex, C.; Triscone, J.M.; Stephan, O. Spectroscopic mapping of local structural distortions in ferroelectric PbTiO3/SrTiO3 superlattices at the unit-cell scale. *Phys. Rev. B* **2011**, *84*, 220102. [CrossRef]
67. Chawla, V.; Holec, D.; Mayrhofer, P.H. Stabilization criteria for cubic AlN in TiN/AlN and CrN/AlN bi-layer systems. *J. Phys. D Appl. Phys.* **2013**, *46*, 045305. [CrossRef]
68. Cooper, V.R.; Rabe, K.M. Enhancing piezoelectricity through polarization-strain coupling in ferroelectric superlattices. *Phys. Rev. B* **2009**, *79*, 180101. [CrossRef]
69. Chen, B.; Zhang, Q.; Bernholc, J. Si diffusion in gaas and si-induced interdiffusion in gaas/alas superlattices. *Phys. Rev. B* **1994**, *49*, 2985–2988. [CrossRef]
70. Schmid, U.; Christensen, N.; Cardona, M.; Lukes, F.; Ploog, K. Optical anisotropy in GaAs/AlSs(110) superlattices. *Phys. Rev. B* **1992**, *45*, 3546–3551. [CrossRef]
71. Gibson, Q.D.; Schoop, L.M.; Weber, A.P.; Ji, H.; Nadj-Perge, S.; Drozdov, I.K.; Beidenkopf, H.; Sadowski, J.T.; Fedorov, A.; Yazdani, A.; et al. Termination-dependent topological surface states of the natural superlattice phase Bi_4Se_3. *Phys. Rev. B* **2013**, *88*, 081108. [CrossRef]
72. Park, C.; Chang, K. Structural and electronic-properties of GaP-AlP (001) superlattices. *Phys. Rev. B* **1993**, *47*, 12709–12715. [CrossRef]
73. Romanyuk, O.; Hannappel, T.; Grosse, F. Atomic and electronic structure of GaP/Si(111), GaP/Si(110), and GaP/Si(113) interfaces and superlattices studied by density functional theory. *Phys. Rev. B* **2013**, *88*, 115312. [CrossRef]
74. Abdulsattar, M.A. SiGe superlattice nanocrystal pure and doped with substitutional phosphorus single atom: Density functional theory study. *Superlattices Microstruct.* **2011**, *50*, 377–385. [CrossRef]
75. Botti, S.; Vast, N.; Reining, L.; Olevano, V.; Andreani, L. Ab initio and semiempirical dielectric response of superlattices. *Phys. Rev. B* **2004**, *70*, 045301. [CrossRef]
76. Rondinelli, J.M.; Spaldin, N.A. Electron-lattice instabilities suppress cuprate-like electronic structures in $SrFeO_3/OSrTiO_3$ superlattices. *Phys. Rev. B* **2010**, *81*, 085109. [CrossRef]
77. Giorgetti, M.; della longa, S.; Benfatto, M. EXAFS and XANES simulations of Fe/Co hexacyanoferrate spectra by GNXAS and MXAN. *J. Phys. Conf. Ser.* **2009**, *190*, 012145. [CrossRef]

78. Westre, T.E.; Cicco, A.; Filipponi, A.; Natoli, C.R.; Hedman, B.; Solomon, E.I.; Hodgson, K.O. Using GNXAS, a multiple-scattering EXAFS analysis, for determination of the FeNO angle in FeNO7 complexes. *Phys. B Condens. Matter* **1995**, *208–209*, 137–139. [CrossRef]
79. Momma, K.; Izumi, F. VESTA 3 for three-dimensional visualization of crystal, volumetric and morphology data. *J. Appl. Crystallogr.* **2011**, *44*, 1272–1276. [CrossRef]

© 2018 by the authors. Licensee MDPI, Basel, Switzerland. This article is an open access article distributed under the terms and conditions of the Creative Commons Attribution (CC BY) license (http://creativecommons.org/licenses/by/4.0/).

nanomaterials

MDPI

Article

Importance of Interface in the Coarse-Grained Model of CNT/Epoxy Nanocomposites

Ke Duan [1], Li Li [2], Fei Wang [1], Weishuang Meng [1], Yujin Hu [1,*] and Xuelin Wang [1]

1 State Key Lab of Digital Manufacturing Equipment and Technology, School of Mechanical Science and Engineering, Huazhong University of Science and Technology, Wuhan 430074, China; duank@hust.edu.cn (K.D.); d201880192@hust.edu.cn (F.W.); m201770371@hust.edu.cn (W.M.); wangxl@hust.edu.cn (X.W.)

2 School of Mechanical and Aerospace Engineering, Nanyang Technological University, Nanyang Avenue, Singapore 639798, Singapore; lili_em@hust.edu.cn

* Correspondence: yjhu@mail.hust.edu.cn

Received: 28 August 2019; Accepted: 14 October 2019; Published: 17 October 2019

Abstract: Interface interactions play a crucial role in determining the thermomechanical properties of carbon nanotubes (CNTs)/polymer nanocomposites. They are, however, poorly treated in the current multi-scale coarse-grained (CG) models. To develop suitable CG models of CNTs/polymer nanocomposites, we demonstrate the importance of two aspects for the first time, that is, preserving the interfacial cohesive energy and reproducing the interface load transfer behavior of all-atomistic (AA) systems. Our simulation results indicate that, for CNTs/polymer nanocomposites, the interface cohesive energy and the interface load transfer of CG models are generally inconsistent with their AA counterparts, revealing significant deviations in their predicted mechanical properties. Fortunately, such inconsistency can be "corrected" by phenomenologically adjusting the cohesive interaction strength parameter of the interface LJ potentials in conjunction with choosing a reasonable degree of coarse-graining of incorporated CNTs. We believe that the problem studied here is general for the development of the CG models of nanocomposites, and the proposed strategy used in present work may be applied to polymer nanocomposites reinforced by other nanofillers.

Keywords: interface force fields; CNTs/epoxy nanocomposites; coarse-grained model; molecular dynamics

1. Introduction

During the past few decades, polymer nanocomposites (PNCs) have received significant attention and have long been at the forefront of research in the polymer community for their broad range of potential applications [1–7]. PNCs, different from traditional composite materials, exhibit quite complex thermomechanical properties which are strongly related to the physics and chemistry of the inclusion of nanofillers, including nanoparticles, carbon nanotubes (CNTs) [4,8], and graphenes [9,10]. Among all PNCs, the most attractive materials may be those reinforced by carbon nanotubes, due to their practical engineering applications such as aerospace materials, packing, and electromagnetic interference (EMI) shielding [4,11,12].

Although very promising, most of the practical applications of CNTs-reinforced PNCs are currently hindered by the lack of comprehensive understanding of their microstructure-to-property relationships [5,13]. As widely demonstrated, the reinforcing efficiency of CNTs in the mechanical properties of PNCs is highly dependent on several factors such as dispersion state of CNTs [14,15], CNTs waviness and orientations [16], and surface functionalization [17]. In general, it is extremely challenging to control and measure these factors in experiments, and their influences on the mechanical properties of CNTs-reinforced PNCs can only be studied via a variety of computational modeling approaches [18,19]. Overall, these computational methods can be divided into three categories—(1) molecular- or micro-scale methods (such as molecular dynamics (MD) and Monte Carlo simulations), (2) meso-scale methods exemplified by dissipative particle dynamics (DPD) and Brownian dynamics, and (3) macro-scale methods (e.g., equivalent-continuum, and finite element method) [18]. Among these numerical approaches, MD simulations are an extremely powerful tool due to their capability to provide realistic interactions between CNTs and polymer-matrix. The interactions are, however, typically approximated or poorly treated in other methods. Such advantage enables MD simulations to probe interfacial phenomena occurring at nanoscale, such as stick-slip damping mechanism [20], and theoretically to consider any structural characteristics of CNTs (dispersion state, waviness, orientations and so on) for studying the mechanical properties of PNCs. The computational efforts of using all-atomic MD simulations to establish the desired structural-properties relationships of CNTs-reinforced PNCs have, however, been limited by the massive computational resources needed which are typically beyond the ability of modern facilities.

Alternatively, developing coarse-grained (CG) models that are capable of preserving the mechanical properties of their all-atomistic (AA) counterparts has been an attractive approach for providing fundamental insights into the mechanical behaviors of polymers or their nanocomposites [21–23]. In comparison with AA simulations, the accessible time and length scales for CG models are substantially increased by removing the "unessential" atomistic features within molecules and thus significantly reducing the number of degrees of freedom to obtain the properties of interest [24]. Several approaches have been proposed to derive the force fields of CG models from their AA counterparts, such as iterative Boltzmann inversion (IBI) method [25], force matching [26,27], and inverse Monte Carlo methods [28]. As for the CNTs-reinforced PNCs, although the force fields of CNTs and polymer matrix can be successfully derived, respectively, from the force matching and IBI approach [23,29,30], there is still a lack of the definition on the interface interactions between CNTs and the polymer matrix material. To the best of the authors' knowledge, only a few studies are devoted to the development of the CG models of CNTs-reinforced PNCs [21,31,32]. In these studies, the CNTs/polymer interface was simply described by Lennard-Jones (LJ) potentials where the key parameters of interfacial LJ pairs are directly predicted by using an arithmetic mean manner without deep theoretical support. Furthermore, the degree of coarse-graining for CNTs was chosen for the CG models of CNTs-reinforced PNCs without any evaluation on its reasonability [21,32]. As a result, the interfacial characteristics of CNT-reinforced PNCs predicted the AA and CG models might be significantly deviated, leading to unforeseen artifacts in the mechanical properties obtained from the CG models.

In the present study, we aim to provide insights into the problem of developing a reasonable CG model for CNTs-reinforced PNCs. Specific attention is focused on the comparison of the interface characteristics between AA and CG models. We have two key findings—(1) the interfacial cohesive energy across the CNTs/epoxy interface modeled using CG models, which is described by LJ potentials similar to the previous researches, is significantly overestimated compared with its real AA counterpart; and (2) the adopted degree of coarse-graining λ for CNT is found to play a critical role in affecting the interfacial shear behavior, emphasizing the importance of evaluating the reasonability of the adopted λ and the interfacial force fields before using the CG models to probe mechanical properties.

2. Simulation Methods

2.1. Potentials and Molecular Models

Many universal force fields have been developed for describing the interactions of AA models, such as Dreiding [33], consistent valence force field (CVFF) [34], and polymer consistent force field (PCFF) [35]. CG potentials are, however, not unique, even for a specific material. In fact, they are highly dependent on many factors, such as the coarse-graining scheme and the thermodynamic state of CG models (temperature and pressure) [23]. The main purpose of this work is to provide an insight into the development of a reasonable CG model considering carefully the crucial role of interface on the mechanical properties of CNT/epoxy nanocomposites. To achieve this, we chose a typical CNT/epoxy nanocomposite in which (5,5) CNT was considered as the reinforcement, and a commonly used engineering cross-linked system which composed of diglycidyl ether of bisphenol A (DGEBA) and 4,4-diaminodiphenyl methane (DDM) was selected as the matrix.

2.1.1. All-Atomic Model

The initial nanocomposite model, as presented in Figure 1a, was constructed with two procedures. First, the 75.7 Å long (5,5) CNT (periodically in the longitudinal direction) was placed at the center of a simulation unit box. Then, the DGEBA and DDM molecules (there are 574 and 287 molecules for DGEBA and DDM, respectively.) were packed into the extra space of the simulation box with a Monte Carlo style similar to our previous work [2]. The general-purpose Dreiding force field was adopted to describe the interactions between atoms, in combination with long-range Columbic interactions where atomic partial charges were determined by the charge equilibration (QEq) method [33,36]. The used cutoff distance for van der Waals and electrostatic interactions was 12 Å. To relax the initial simulation model, an equilibration procedures similar to our previous work was carried out [23,37]. After that, the fully relaxed nanocomposite model was cross-linked to a specific degree of cross-linking (DOC) at temperature 700 K using a dynamic cross-linking strategy [37]. The details for the adopted dynamics cross-linking strategy can be found in our previous paper [37]. Finally, the cross-linked nanocomposite model was again equilibrated to a target temperature. For more details about the constructed model and equilibration procedures, please refer to Supplementary Section S1.1. Periodic boundary conditions were imposed on all three directions of the simulation cell, namely, the x, y, and z directions. A timestep 1 fs was used for all AA simulations.

2.1.2. Coarse-Grained Model

The construction and equilibration processes of the CG models for CNT/epoxy nanocomposites are similar to those of their AA counterparts and we present these technical details in Supplementary Section S1.2. In this section, we are mainly concerned with the issue of force fields used in the CG model of CNT/epoxy nanocomposites. For the CNT/epoxy nanocomposites, the force fields can be divided into three parts—(1) reinforcement force fields, that is, the interactions for particles in one (5,5) CNT or between different (5,5) CNTs, (2) matrix force fields, namely both the intra- and inter-molecule interactions for DGEBA and DDM molecules, (3) interfacial force fields between the (5,5) CNT and the cross-linked epoxy matrix.

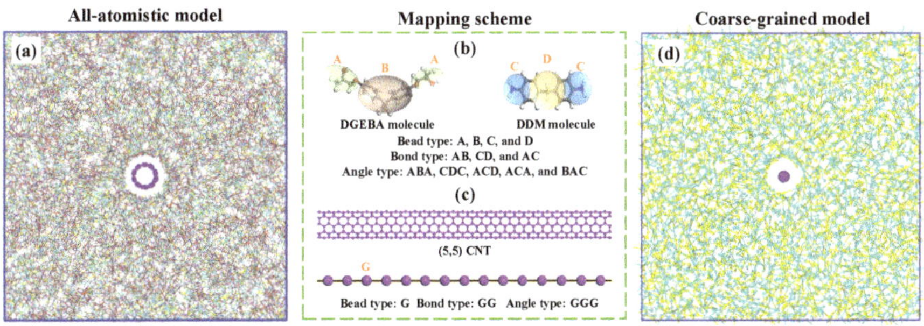

Figure 1. Schematic view of mapping all-atomistic model to coarse-grained model for epoxy nanocomposites reinforced by (5,5) carbon nanotube (CNT). (**a**) Snapshot of the all-atomistic model, (**b**) the adopted mapping scheme for DGEBA and DDM molecules, (**c**) the adopted mapping scheme for (5,5) CNT, and (**d**) snapshot of the resulting coarse-grained model.

For CNTs, the fundamental mechanical parameters of their CG models can be derived from AA simulations based on the matching of energies and mechanical features, including tensile stiffness, bending stiffness and adhesion properties. Specifically, the force field of (5,5) CNT used here has been already reported by Buehler [29] and was widely-adopted for investigating the tensile and viscoelastic properties of CNT-based materials [38–40]. As for the cross-linked epoxy matrix studied here, the coarse-grained force fields were determined through a combination of the IBI method and the machine-learning based approach, as reported in our previous work [23]. The mapping scheme for DGEBA and DDM molecules are the same as that in our previous work [23], illustrated in Figure 1b. The IBI method was used to preserve the structural aspect of the matrix molecule and thus determine the bonded interaction parameters (including the bond and angle potentials). By using a robust machine-learning based technique, we determined the nonbonded interaction (the 12-6 LJ potential form) to reproduce the density, glass transition temperature and elastic moduli of cross-linked epoxy at a wide range of temperatures and various values of DOC. The main feature of the nonbonded potential is that the strength parameter of the cohesive interaction ϵ was DOC- and temperature-dependent. The CG model for cross-linked epoxy has four types of CG beads (beads 'A', 'B', 'C', and 'D') and the detailed force fields are presented in Supplementary Section S2.1.

Since the CG potentials of the CNT and cross-linked epoxy were derived separately by means of totally different approaches, the determination of the CG potentials of CNT/epoxy nanocomposites requires an additional definition of the CNT/epoxy interface interaction. Beside, the degree of coarse-graining λ for CNT may have a significant effect on the interfacial interactions and thus on the mechanical properties of the nanocomposites due to the fact that a large λ typically leads to artificial surface roughness [41]. However, to the best of our knowledge, there is nearly no previous literature considering the crucial issue whether the adopted degree of coarse-graining λ for CNT is reasonable for studying the problems at hand?

With the above question in mind, the CG models of the CNT/epoxy nanocomposites were constructed in such a way that the degree of coarse-graining λ can be adjusted from 4 Å to 10 Å. Figure 1c depicts the mapping scheme of (5,5) CNT using λ = 5 Å. The bead within a single CNT was labeled as 'G', forming 1 bond type 'G G' and 1 angle type 'G G G'. The force fields of (5,5) CNT with various values of λ can be found in Supplementary Section S2.2. Based on the mapping rules described above, a snapshot of the CG models of CNT/epoxy nanocomposites, in which epoxies are not cross-linked yet, is shown in Figure 1d. For all CG simulations, timestep is 4 fs unless otherwise mentioned. All the CG and AA simulations were performed with large-scale atomic/molecular massively parallel simulator (LAMMPS) [42] and visualized via the Open Visualization Tool (OVITO) [43].

2.2. Evaluation of the Interface Reasonability

As is widely-known, the key feature of nano-reinforcements is the extremely larger contact surface area with polymer-matrix in comparison with traditional fillers, enabling enormous interesting properties of the resulted nanocomposites [5,44]. Therefore, the load transfer behavior of the CNT/epoxy interface should be well preserved under coarse-graining, otherwise the developed CG models may lead to unforeseen artifacts on the studied properties. Essentially, the load transfer of a CNT/epoxy interface can be characterized by the force-separation responses in radial opening and axial sliding mode separations [45–47]. Similarly, we here use this approach to evaluate the interface behavior of the coarse-grained CNT/epoxy interface which is compared to their AA counterpart. Specifically, we are more concerned with the equivalence issue of CNT/epoxy interface between the AA and CG models. To accompany this, we adopt a strategy called as "monomeric separation response" which is done by performing radial opening and sliding mode separation simulations of a single matrix molecule onto the CNT surface and recording the force-separation response, as illustrated in Section 2.2.2. As demonstrated by previous research, the interfacial strength of the CNT/polymer interface is typically an additive effect of the monomeric friction [48]. Thus, it is feasible to use the "monomeric separation response" approach for evaluating the interface load transfer behavior for the CNT/epoxy nanocomposites here.

2.2.1. Calculation of the Interfacial Cohesive Energy

Before illustrating the strategy of "monomeric separation response," it is necessary to pay attention to the interface cohesive energy, which determines the traction force significantly for both radial opening and axial sliding separation. Thus, the first thing to be satisfied is an identical interfacial cohesive energy between AA and CG models, otherwise the evaluation for load transfer of the CNT/epoxy interface has no meaning.

The total energy of a CNT/epoxy nanocomposite unit cell can be expressed as

$$E_{total} = E_{CNT} + E_{epoxy} + E_{interface} \tag{1}$$

where E_{total} indicates the total energy of the simulation system, E_{CNT} represents the energy of embedded CNTs, E_{epoxy} is the energy of the epoxy matrix, and $E_{interface}$ represents the energy of the CNT/epoxy interface. The interfacial cohesive energy per unit length can then be calculated as [49]

$$\gamma = \frac{E_{\text{interface}}}{2A} \tag{2}$$

where A is the contact area of the interface, determined as $A = \pi dL$ where d and L are the diameter (6.78 Å) and the embedded length of the used (5,5) CNT, respectively. It should be mentioned that the value of L is determined to be the dimension of simulation box at a full equilibrated state along the longitudinal direction of the (5,5) CNT.

2.2.2. Characterization of the Interface Load Transfer

The behavior of the CNT/epoxy interface is evaluated by the force-separation responses in axial opening and sliding mode separations, as shown in Figure 2. For the AA system, as depicted in Figure 2a, a fully equilibrated DGEBA molecule was forced to separate from the (5,5) CNT along the radial direction for the radial opening mode, and to slide on the CNT surface along its axial direction for the sliding mode separation. For the CG models, the simulation procedures are the same, as done for their AA counterparts. As mentioned in Section 2.1.2, λ can be changed and then leads to totally different CG force fields. Thus, it is of great significance to examine whether λ has any effects on the interface behavior. In this work, 7 different λ values ranging from 4 to 10 Å were considered. Figure 2b,c illustrated the schematic view of "monomeric separation response" for CG interface models with λ being 5 and 10 Å, respectively.

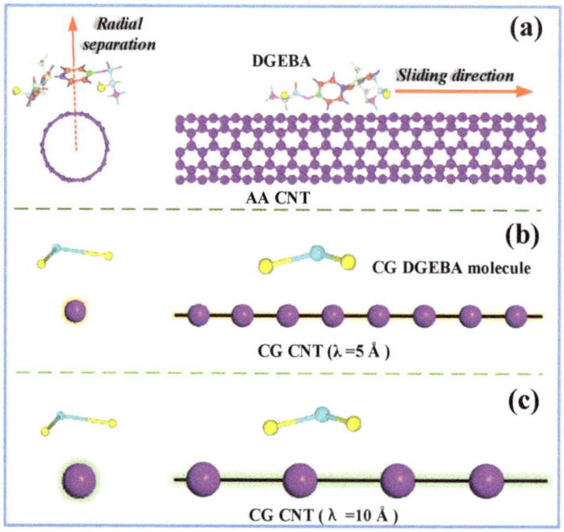

Figure 2. Illustration of adopted strategy "monomer separation response" for evaluating the interface load transfer of CNT/epoxy nanocomposites: (**a**) all-atomistic (AA) system, (**b**) coarse grained (CG) model with λ = 5 Å, and (**c**) CG model with λ = 10 Å. The "monomer separation response" includes radial opening simulation which is done by separating a DGEBA molecule from the (5,5) CNT along its radial direction, and axial sliding mode separation simulation which is done by forcing the DGEBA molecule slides on the CNT surface along its axial direction.

3. Results And Discussion

3.1. Preserve the Interface Cohesive Energies

As described in Section 2.2.1, the interface cohesive energy is a dominant factor in affecting the load transfer across the interface. Thus, the AA interface cohesive energy between CNT and epoxy matrix should be examined and preserved before evaluating the equivalence issue of the interface load transfer between the AA and CG models. To achieve this goal, we constructed 8 simulation models for the CNT/epoxy nanocomposites, including 1 AA (Figure 1a) and 7 CG models (an example is shown in Figure 1d). The degree of cross-linking for all these 8 models are kept the same value of 90%. The target temperature to extract the interface cohesive energy of various simulation models was 300 K. The CNT is periodical along its longitudinal direction to eliminate the end edge interaction effects, in order to determine an accurate value of the interface cohesive energy.

Figure 3a shows the comparison of the interface cohesive energy for both the AA (red dash line) model and the CG models with different values of λ. The obtained interface cohesive energy for all the CG models is not equal to that of the AA model. In fact, the interface cohesive energy of any CG model is significantly higher than that of its AA counterpart. This phenomenon strongly emphasizes the fact that well description of the interactions for each constitutive law of the phases in nanocomposites does not imply the correctness of the whole nanocomposite model. In other words, the interface interactions between two components may be significantly altered when developing a coarse-grained model of a AA system. Another thing can be observed from Figure 3 is that there is no apparent trend between the interfacial energy (CG models) and the degree of coarse-graining λ. This is mainly due to the reason that the interaction cohesive strength ϵ increases linearly (see black line of Figure 3b with the degree of coarse-graining λ, however, the number of vdWl pairs decrease linearly with λ (under the ideal conditions). For instance, Figure 2c presents the condition where the λ is two times that of the condition shown in Figure 2b. Since the length scale parameter σ (see Equation (3)) is independent of λ, the distributed state of matrix beads is expected to be the same for the two cases under ideal conditions. Thus, the number of vdWl pairs for condition shown in Figure 2c is only one half that for condition shown in Figure 2b. Considering the fact that the energy of single vdWl pair for Figure 2c is two times that of for Figure 2a, we can conclude that the interfacial energy for the two cases is the same under ideal conditions. However, for the real CG simulations the distribution state of matrix beads will be affected by not only the initial sampling of the matrix molecules but also the varied λ (not significantly), leading to the fluctuation of interfacial energy of CG models at different degree of coarse-graining λ.

To understand the observed inconsistency of interface cohesive energy between AA and CG models, the key factor to be considered is the used nonbonded potentials for describing the interface. For the nanocomposites studied here, the nonbonded interactions for both AA and CG models are modeled using 12-6 LJ potential (as mentioned in Section 2), having an energy expression:

$$E_{LJ} = 4\varepsilon\left[\left(\frac{\sigma}{r}\right)^{12} - \left(\frac{\sigma}{r}\right)^{6}\right] \tag{3}$$

where r represents the distance between AA atoms or CG particles, ε indicates the depth of the potential well, and σ indicates the radial distance at which the inter-particle potential energy is zero. The detailed values of ε and σ for epoxy and CNT particles are presented in Supplementary Section S2. As for the interface cohesive energy of CG models, they are contributed by 4 interface LJ pairs, that is, GA, GB, GC, and GD. For a traditional manner, the parameters of these 4 LJ pairs are typically determined by the following forms, as done by previous papers [21,31,32].

$$\varepsilon_{ij} = \sqrt{\varepsilon_{ii}\varepsilon_{jj}}, \qquad \sigma_{ij} = (\sigma_{ii} + \sigma_{jj})/2 \tag{4}$$

in which subscript i represents the CNT particle 'G', and subscript j denotes different particle types of epoxy (particle type A, B, C, and D). In this manner, the cohesive interaction strength parameters $\varepsilon_{interface}$ as a function of λ of these CG models are shown in the black line of Figure 3b. It should be clarified here that the value of ε_{ij} is identical for the 4 LJ pairs (GA, GB, GC, and GD) which contribute to all the interface van der Waals (vdWl) energies and we name this unified value $\varepsilon_{interface}$. This is because all epoxies particle types (particle type A, B, C, and D, as shown in Figure 1b share an identical value of ε (we refer the reader to our previous paper [23] for detailed reasons).

Since both $\varepsilon_{interface}$ and σ of the interfacial LJ interactions are critical for contributing to the interface cohesive energy of the CG models, we separately examined whether their characteristics are well captured by the CG models. We proceeded to evaluate the reasonability of length scale parameter σ of the 4 interface LJ pairs (σ_{GA}, σ_{GB}, σ_{GC}, and σ_{GD}) for CG models. This is done by comparing the interfacial radial distribution function (RDF) curves for the AA and CG models. Note that the RDF curves characterize the probability of locating epoxy molecules in a surface of CNT at the distance r, and thus characterize the situation of the effective zone of the particles that contribute to the interfacial cohesive energy. We selected a maximum r = 30 Å and divided it into 60 small sections to determine the RDF curves (please refer to Supplementary Section S3 for more details). It should be noted that we chosen the maximum r to be 30 Å because this distance is equal to the cutoff distance for the van der Waals force of the CG models. Beyond this distance, the interactions between CNT and epoxy molecules are negligible. The comparison of the RDF curves for the four types of matrix beads between the AA and CG models is shown in Figure 4. It can be observed that, although the RDF curves for both the AA and CG models exhibit relatively high fluctuation, their RDF curves are generally in good agreement with each other. The relatively high fluctuation is mainly attributed to the lack of movement of matrix particles because of the high degree of cross-linking (DOC) and the glassy state (the glass transition temperature for DOC-90 cross-linked epoxy is ~437.32 K) [23]. The consistent RDF curves imply that the distributing state of epoxy molecules around the CNT surface of the AA model can be faithfully described by the CG model in which the length scale parameter σ of the 4 interface LJ pairs are determined by Equation (4). Thus, the overestimated interface cohesive energies in the CG models mainly originate from the cohesive interaction strength parameter $\varepsilon_{interface}$ used for modeling the interfacial LJ interactions.

To preserve the interface cohesive energy of the AA models, we phenomenologically adjust $\varepsilon_{interface}$ for CG models with different values of λ. The adjusted value of $\varepsilon_{interface}$ as a function of λ is shown in the red line of Figure 3b. It is clear that $\varepsilon_{interface}$, which preserves the AA interface cohesive energy for CG models, is much smaller than that predicted by traditional model (black line) using Equation (4), emphasizing the importance of carefully re-adjusting the interface interactions when modeling the nanocomposites via CG models. The trend of $\epsilon_{interface}$ with the degree of coarse-graining λ can be fit with a continuous quadratic polynomial function of the form $\varepsilon_{interface} = a\lambda^2 + b\lambda + c$, where a = -0.02075 kcal/molÅ^2, b = 0.556 kcal/molÅ, and c = -0.45265 kcal/mol. This relationship can be used to determine the value of $\epsilon_{interface}$ for a specific adopted λ within the range of 4 to 10 Å which has not given explicitly by our simulations.

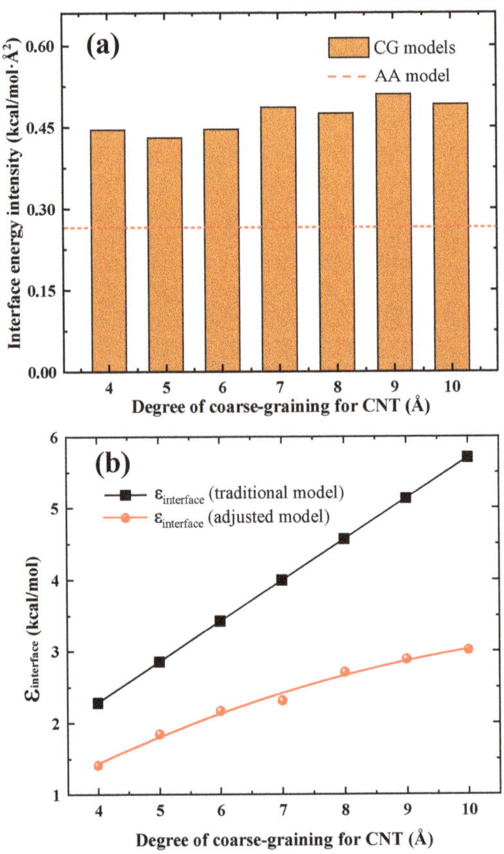

Figure 3. (**a**) Comparison of the interface cohesive energies between the AA and CG models, and (**b**) $\varepsilon_{interface}$ vs. λ curves for the traditional interface model (black line) and the adjusted interface mod el (red line) in this work.

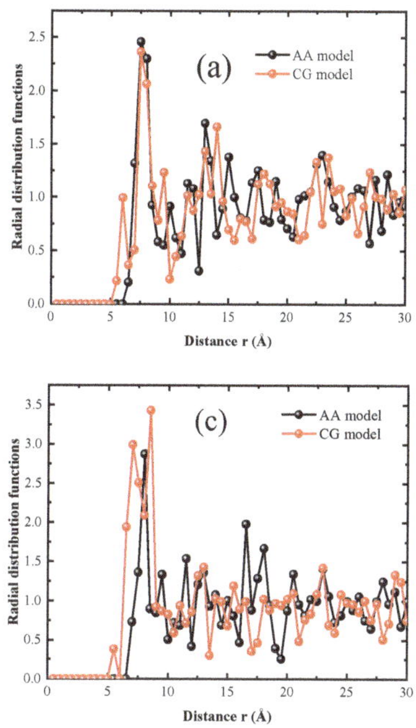

Figure 4. Radial distribution function (RDF) curves for (**a**) particle type 'A', (**b**) particle type 'B', (**c**) particle type 'C', and (**d**) particle type 'D'.

3.2. *Effects of λ on the Interfacial Load Transfer*

In Section 3.1, we preserved the interfacial cohesive energy of the AA models by phenomenologically adjusting the cohesive interaction strength parameter $\varepsilon_{interface}$ of the CG models. In this section, we turn our focus on evaluating the equivalence issue of interface load transfer between the AA and CG models using the "monomeric separation response" approach. It should be noted that the interface energy of the CG models in this section is consistent with their AA counterparts by using the phenomenologically determined $\epsilon_{interface}$ in Section 3.1 to adjust the interface cohesive energy of the CG models.

As described in Section 2.2.2, 7 different values of λ were considered. Thus, there are totally 8 interface models including 1 AA and 7 CG interface models. For each model, a DGEBA molecule was fully equilibrated on the CNT surface and subjected to radial and sliding separations to obtain the force-separation responses using two separation simulations.

For the sake of convenience, we designated these CG interface models in terms of "λ-d" terminology where d is equal to the equilibrium distance between two CNT beads, characterizing the degree of coarse-graining, and indicates the actual value of λ.

3.2.1. Separation Force Due to Radial Opening Mode

Figure 5a illustrates the traction force of the DGEBA molecule for both the AA (black solid line) interface model and the 7 CG interface models (dash lines). All the force-separation curves exhibit a similar pattern in which the traction force initially increases rapidly to peak force and then gradually decreases to zero with the increasing separation displacement, which is typical for vdWl-dominated interface [45,46]. Surprisingly, the peak traction force for all these CG interface models is, however, much smaller than that of their AA counterparts, implying that the interface load transfer in the CNT radial directions is underestimated by the CG models. As observed from Figure 5a, the peak traction force for the CG models is around 0.28 nN, which is less than one thirds that of the AA model (∼1.03 nN). The force inconsistency between the AA and CG models indicates that the CG model cannot simultaneously preserve both the force (radial traction force here and sliding force in Section 3.2.2) and the interfacial energy, which has also been observed in other literature [50,51]. This phenomenon is mainly due to the shift between enthalpy and entropy upon coarse-graining AA models [50–52]. As for these CG models, no evident dependence is observed between the studied λ and the resulted radial traction forces.

3.2.2. Sliding Force Due to Axial Sliding Mode

Force-separation responses for the axial sliding mode are shown in Figure 5b, all of which exhibit a sinusoidal-alike pattern which arises from the periodic crystal structure of both the AA and CG CNT models along their longitudinal direction. The periodic distance is equal to λ for the CG models (Figure 2b,c), while is the distance between two neighboring hexagonal lattice centers along the longitudinal direction for AA CNT (∼2.5 Å, Figure 2a). For instance, the red dash curve in Figure 5b shows a periodic 10 Å for the λ-10 CG interface model.

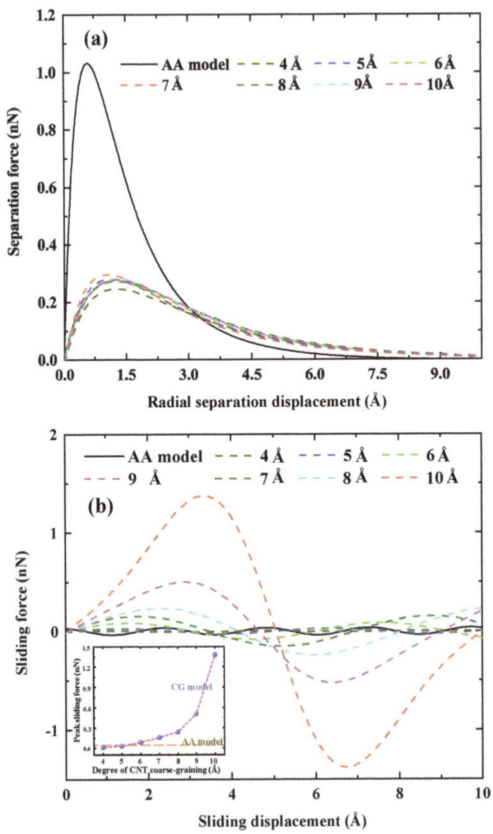

Figure 5. Force-separation curves of the "monomeric separation response" for the AA and CG interface models, (**a**) radial opening mode, and (**b**) axial sliding mode.

Aside from the sinusoidal-alike pattern of these force-separation curves, a very important observation in Figure 5b is the difference in peak sliding forces between the AA and CG models.

As clearly illustrated in Figure 5b, the peak sliding force of CG models is strongly dependent on the adopted λ, exhibiting a nonlinear fashion as demonstrated by the inset of Figure 5b. Specifically, the peak sliding force increases slowly when the adopted λ is less than 8 Å, whereas shows a drastic increase when λ is beyond 8 Å. As a comparison, the required peak sliding force is 1.37 nN for the λ-10 interface model, which is almost 3 orders of magnitude higher than that of the λ-4 model (0.0054 nN). The real peak sliding force derived from the AA model is about 0.0354 nN (the black solid line in Figure 5b), which is much less than that of the studied CG models. From the sliding mode separation characteristic point of view, we can conclude that in order to capture the AA sliding characteristics reasonably, there is a limit of the degree of coarse-graining λ of the CNT beyond which the mechanical properties of the CG model of the CNT are too deteriorated. As seen, the interface appears to be too rough if λ is too large and too smooth if λ is too small. Here, the suitable λ of (5,5) CNT is around 5 Å due to the reason that the peak sliding force of the λ-5 interface model (0.026 nN) is the closest to that of the AA model (0.0354 nN), as suggested by the inset of Figure 5b.

3.3. *Effect of λ on Elastic Moduli*

As illustrated in Section 3.2, the interface load transfer behavior of CG models is typically different to their AA counterparts and is, in fact, highly dependent on λ. As a result, implementing different values of λ into the CG models may lead to seriously unforeseen artifacts on the predicted mechanical properties of the CNT/epoxy nanocomposites. To justify this, we investigated a simple but representative problem right here for CNT/epoxy nanocomposites, namely the effect of the CNT's length on the tensile moduli of CNT/epoxy nanocomposites.

The CG models were constructed in such a way that 4 parallel coarse-grained (5,5) CNT with their axes along the same direction (which is called the longitudinal direction of the nanocomposites) were fully embedded into the epoxy-matrix. Furthermore, we name the other two directions as the transverse (or lateral) directions.

Figure 6 shows parts of the fully relaxed configurations of the CG models of epoxy-matrix nanocomposites, each of whose cells has 4 embedded CNTs. When modeling these CNT/epoxy nanocomposites, 4 different values of λ were considered, that is, λ = 4, 5, 6 and 10 Å. For each studied λ, 6 different length-to-diameter ratios, ranging from 10 to 150, were considered.

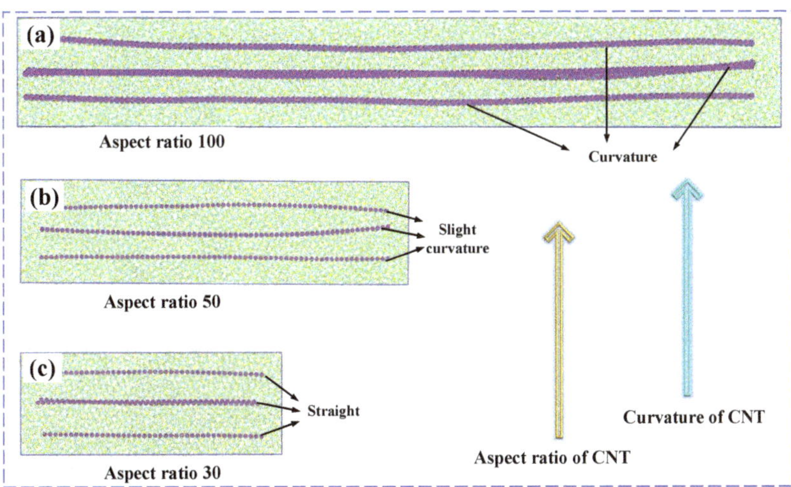

Figure 6. Fully relaxed configurations of the CG models of epoxy-matrix nanocomposites, each of whose cells have 4 embedded CNTs, (**a**) the aspect ratio of CNTs is 100, (**b**) the aspect ratio of CNTs is 50, and (**c**) the aspect ratio of CNTs is 30. All the degrees of coarse-graining of the CNTs are chosen as $\lambda = 5$ Å.

To differentiate these constructed CG models, we use a terminology of "λ-d-α" where d indicates the actual value of the degree of coarse-graining of the CNT, as defined above, and α represents the specific value of the aspect ratio of the embedded CNT. For all the CG models, the mass fraction of CNTs was kept the same at 5% to eliminate the influence of CNT contents.

We proceeded to evaluate the influence of λ on the elastic responses of nanocomposites in the two transverse directions. Figure 7 shows the comparison of the averaged lateral elastic modulus for these CG models. It is clear that the lateral elastic modulus for all these CG models are comparable to each other, yielding values of around 4.2 GPa. Specifically, the value is also comparable to the elastic modulus of the epoxy matrix (~3.97 GPa), implying that the embedded CNTs show a negligible improvement on the lateral elastic modulus of the CNT/epoxy nanocomposites. This phenomenon is consistent with the reported AA simulation results in which limited improvement or even deterioration were observed on the transverse elastic modulus of CNT/epoxy nanocomposites [44].

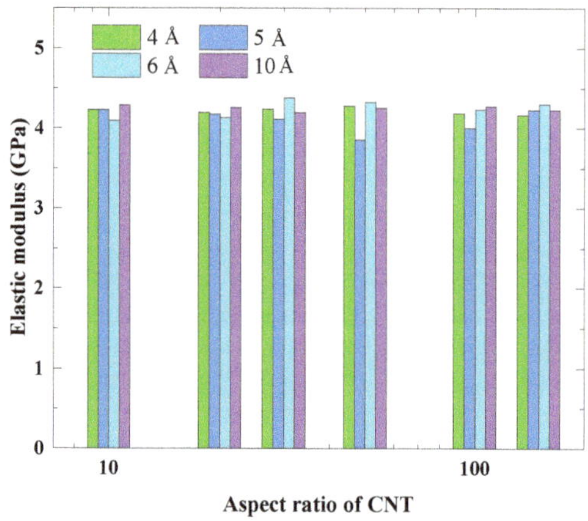

Figure 7. Averaged lateral elastic moduli of epoxy-matrix nanocomposites reinforced by the CNTs which have different aspect ratios and various degrees of coarse-graining.

In Section 3.2.1, although the radial opening mode simulations show that the interface load transfer predicted by the CG models was underestimated, this deviation shows a negligible effect on the lateral elastic moduli. Since the aim of developing CG models is to reduce the complexity of their AA counterparts while retaining all the most important features, we also chose $\lambda = 5$ Å. There is a price we have to pay for developing an efficient but simple CG model.

We next evaluate the tensile responses along the longitudinal (axial) direction of the embedded CNTs, which are illustrated in Figure 8. Different from what we observed in Figure 7, the adopted λ has a significant effect on the axial elastic modulus (E_{axial}) of the CNT/epoxy nanocomposites. As shown in Figure 8, for each fixed λ value, E_{axial} increases with the increasing CNT aspect ratio in a nonlinear fashion and tends to be plateau values in the high aspect ratio regimes. In particular, the critical aspect ratio where E_{axial} starts entering the plateau stage appears to be larger for higher λ. For instance, the critical aspect ratio increases from 50 to 100 when λ increases from 5 to 6 Å. When λ is increased to 10 Å, the measured E_{axial} is even not converged at the aspect ratio of 150 (CG models with a larger aspect ratio of the CNT were not carried out for the sake of computational resource). For each studied aspect ratio, the measured E_{axial} of the CG models increases with the increasing value of λ and the deviation become larger at the cases of higher CNT aspect ratios. For instance, the calculated E_{axial} for λ-10-10 (4.725 GPa) is only slightly larger than that of λ-4-10 (4.427 GPa). However, the E_{axial} of λ-10-150 (14 GPa) is almost three times that of λ-4-150 (5.83 GPa). This phenomenon is attributed to the higher interface load transfer along the longitudinal direction when a larger λ was adopted. The better interface load transfer ability generally enables higher stress transfer from the matrix to CNT and thus leads to a larger tensile modulus of the CNT/epoxy nanocomposites [53,54]. As expected, E_{axial} of CG models will be further increased if considering a larger λ.

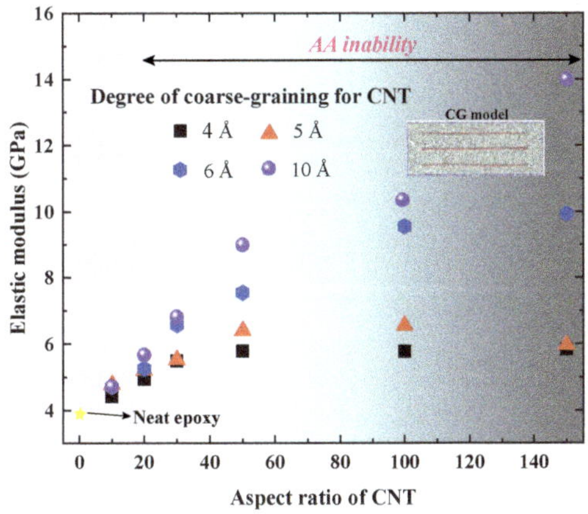

Figure 8. Axial elastic moduli of epoxy-matrix nanocomposites reinforced by the CNTs which have different aspect ratios and various degrees of coarse-graining.

The astonishing observations from Figure 8 strongly emphasize the importance of adopted λ in determining the studied tensile behavior of CNT/epoxy nanocomposites. According to the "monomeric separation response" analysis shown in Section 3.2, λ = 5 Å might be the most successful of the many values to the reproduction of the AA interface load transfer. As a result, the nanocomposites CG models with λ = 5 Å appears to be a suitable choice among the studied cases to capture the actual mechanical properties of CNT/epoxy nanocomposites.

Prior to this work, the most successful approaches in characterizing the CNT reinforcing efficiency in terms of different CNTs lengths use finite element analysis (FEA) or multi-scale modeling (such as the shear lag model, Halpin-Tsai, Mori-Tanaka) [53–57]. The tensile modulus predicted by these approaches typically converges almost to the rule of mixtures, which is much more larger than that calculated by our CG models [53]. This phenomenon is attributed to the unrealistic interactions of the CNT/matrix interface for these multi-scale modeling techniques which do not capture the weakly nonbonded vdWl nature of the interface. Moreover, CNTs with a high aspect ratio are assumed to be straight in both FEA and multi-scale modeling, which is not the real situation in the synthesized CNT/epoxy nanocomposites [58]. In our CG simulations, despite the momentum control of each embedded CNT (done by using the *fix momentum* command in LAMMPS), the curvature of embedded CNTs can also be observed clearly (details are shown in Supplementary Section S4). As widely demonstrated by previous studies, the waviness of CNTs has an evident deterioration on the tensile modulus of nanocomposites [37,56] and thus leads to a smaller critical CNT aspect ratio where the convergence of tensile modulus occurs.

3.4. Determination of the Interface Force Fields

In previous sections, we have demonstrated the critical role of λ in affecting the CNT/epoxy interface load transfer capability and the determined mechanical property of CNT/epoxy nanocomposites. In the principle of preserving the actual (AA model) interface behavior, λ = 5 Å appears to be reasonable in developing the CG models for CNT/epoxy nanocomposites. For CG models using λ = 5 Å, the cohesive interaction strength parameter $\varepsilon_{interface}$ is determined to be 1.849 kcal/mol. This value is, however, determined for the CNT/epoxy nanocomposites whose DOC of epoxy matrix is 90%. Besides, the simulation temperature is 300 K. Thus, whether the LJ parameters derived from DOC-90 and temperature 300 K would work at other DOC and temperatures remains to be validated.

To answer this question, we proceeded to evaluate the temperature issue of the developed CG models. As concluded from the previous sections, for a fixed λ, the interface load transfer ability is solely dependent on the interface cohesive energy. In other word, the evaluation of whether the CG force fields are able to work at other temperatures can be done through comparing the interface cohesive energy between the AA and CG models at various simulation temperatures. Figure 9 shows the comparison of interface cohesive energy between AA and CG models where the DOC is 90%. As temperature increases, the interface cohesive energy decreases for both AA and CG models. This phenomenon is originated from the reduced number of matrix molecules surrounding the CNT surface at higher temperatures because the matrix density generally decreases as temperature increases. Moreover, the deviation of interface cohesive energy between the AA and CG models at each studied temperature is slight (relative error less than 2%), indicating that the developed CG force fields of CNT/epoxy nanocomposites generally work well over a wide temperature range. It should be stressed here that the CG models of CNT/epoxy nanocomposites which performed well at different temperatures is a "set" of CG potentials because the potentials for cross-linked epoxies change at varies temperatures (see Supplementary Section S2.1 for details).

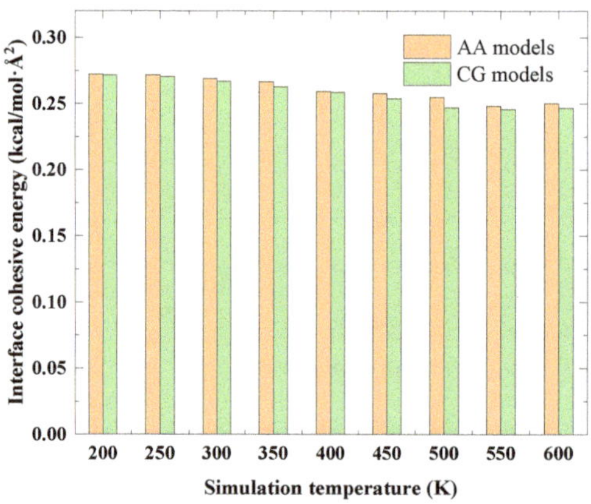

Figure 9. The developed CG models for CNT/epoxy nanocomposites work well over a wide temperature range.

We next examine the DOC issue for the CG force fields through considering another 3 different matrix-DOC, that is, DOC-60, DOC-70, and DOC-80. The simulation temperature was kept at 300 K. Table 1 lists the values of the interface cohesive energy determined by the AA and CG simulations for the studied nanocomposites with 4 different values of DOC. The first thing we can observe is that, for AA models, the interface cohesive energy at various DOC are almost identical, which is mainly attributed to the insignificant effects of DOC on the density of the CNT/epoxy nanocomposites (1.198 g/cm^3 for DOC-60 nanocomposites, and 1.206 g/cm^3 for DOC-90 nanocomposites). However, for CG models, the interface cohesive energy was underestimated for DOC-60, DOC-70, and DOC-80 by the CG force fields derived from DOC-90. This observation strongly emphasizes that the CG force fields derived from DOC-90 are unable to transfer to other DOC. Different from the AA models, the DOC of matrix exhibits much more significant influence on the density (1.078 and 1.211 g/cm^3 for DOC-60 and DOC-90 cross-linked epoxy, respectively) [23]. However, this effect on the interface cohesive energy of the nanocomposites CG models could be compensated for by adjusting the cohesive interaction strength parameter of the interface LJ potentials (see Section 3.1 for details). By phenomenologically preserving the AA interface cohesive energy, we determined a DOC-dependent $\varepsilon_{interface}$ for the interface LJ potentials of the CG models, yielding values of 2.331, 2.167, and 2.019 kcal/mol for

DOC-60, DOC-70, and DOC-80 CNT/epoxy nanocomposites, respectively. The adjusted $\varepsilon_{interface}$ leads to a linear variation with the change of DOC, following the expression:

$$\varepsilon_{interface} = 3.287 - 1.594 \times \text{DOC} \tag{5}$$

Table 1. Interface cohesive energy of CNT/epoxy nanocomposites using the AA and CG models.

	DOC-60	DOC-70	DOC-80	DOC-90
AA models	0.272	0.271	0.270	0.269
CG models	0.219	0.229	0.242	0.269
CG-adjusted	0.272	0.271	0.270	0.269

Using these adjusted $\varepsilon_{interface}$, the interface cohesive energy of CNT/epoxy nanocomposites at various values of DOC are well preserved, as listed in Table 1.

After showing that the CG models for CNT/epoxy nanocomposites can also work well at different temperatures and DOC through using DOC- and temperature-dependent cohesive interaction strength ϵ for both epoxy matrix and the CNT/epoxy interface interactions, we provide here some particular notes for the practical use of the developed CG potentials.

- For the reinforcement force fields, the selection of degree of coarse-graining for (5,5) CNT should be ~5 Å, which provides a faithful approximation to the interface load transfer characteristics of the AA models (see potential parameters for CNT (λ = 5 Å) in Supplementary Table S2).
- The cross-linked epoxy matrix force fields are based on our previous work and are shown in Supplementary Table S1 [23].
- The interface force field is modeled using the 12-6 LJ potentials where the cohesive interaction strength parameter ε is determined by Equation (5) (case of λ = 5 Å), whereas the length scale parameter σ is calculated as Equation (4).

4. Conclusions

In summary, we demonstrated for the first time that the interface characteristics of CG models for the CNTs/epoxy nanocomposites are generally inconsistent with their AA counterparts, causing artifact to the predicted mechanical properties. On one hand, the interface cohesive energies of the CG models are significantly higher than that of their AA counterparts, which will inevitably lead to an overestimated interface load transfer ability. On the other hand, the degree of coarse-graining λ of CNT plays a crucial role in affecting the interface load transfer behavior of the CG models. In particular, the peak sliding force for a single DGEBA molecule onto the CNT surface with λ = 10 Å can reach 1.37 nN, which is nearly 2 orders of magnitude higher than that of the real situation derived from the AA model (0.0354 nN). As a result, the predicted mechanical property of nanocomposites shows a very large deviation when various values of λ were adopted for developing the CG models.

To correctly reproduce the real interface (the AA system), our data strongly advocate two design criteria for coarse-grain modeling the CNTs/epoxy nanocomposites—(1) phenomenologically adjusting the cohesive interaction strength parameter ($\varepsilon_{interface}$) of the interface LJ potentials to preserve the interface cohesive energies of AA models and (2) carefully choosing a degree of coarse-graining for the embedded CNTs to capture the actual interface load transfer behavior derived from the AA models. We emphasize that only in this manner are the CG models of nanocomposites capable of providing useful insights into the predicted mechanical properties or mechanisms.

Supplementary Materials: The following are available online at http://www.mdpi.com/2079-4991/9/10/1479/s1.

Author Contributions: K.D.: conceptualization, formal analysis, writing—original draft preparation, writing-review and editing; L.L.: writing—original draft preparation; F.W.: simulation; W.M.: simulation; Y.H.: supervision, project administration, funding acquisition; X.W.: supervision, project administration.

Funding: This research was funded by the National Natural Science Foundation of China (Grant Nos. 51605172 and 51775201), the Natural Science Foundation of Hubei Province of China (Grant No. 2016CFB191), and the Fundamental Research Funds for the Central Universities (Grant No. 2015MS014).

Acknowledgments: Computational resources from SCTS/CGCL HPCC of Huazhong University of Science and Technology and the National Supercomputer Center in Lvliang and Guangzhou of China are gratefully acknowledged.

Conflicts of Interest: The authors declare no conflict of interest.

Data Availability: The data that support our studies is available from the corresponding authors on reasonable request.

References

1. Andropova, U.; Tebeneva, N.; Serenko, O.; Tarasenkov, A.; Buzin, M.; Shaposhnikova, V.; Muzafarov, A. Nanocomposites based on polyarylene ether ketones from sol–gel process: Characterizations and prospect applications. *Mater. Des.* **2018**, *160*, 1052–1058. [CrossRef]
2. Duan, K.; Li, Y.; Li, L.; Hu, Y.; Wang, X. Pillared graphene as excellent reinforcement for polymer-based nanocomposites. *Mater. Des.* **2018**, *147*, 11–18. [CrossRef]
3. Ding, J.; Zhang, H.P.; Li, X.; Tang, Y.; Yang, G. Crosslinked carbon nanofiber films with hierarchical pores as flexible electrodes for high performance supercapacitors. *Mater. Des.* **2018**, *141*, 17–25. [CrossRef]
4. Martin-Gallego, M.; Yuste-Sanchez, V.; Sanchez-Hidalgo, R.; Verdejo, R.; Lopez-Manchado, M.A. Epoxy nanocomposites filled with carbon nanoparticles. *Chem. Rec.* **2018**, *18*, 928–939. [CrossRef]
5. Kumar, S.K.; Benicewicz, B.C.; Vaia, R.A.; Winey, K.I. 50th anniversary perspective: Are polymer nanocomposites practical for applications? *Macromolecules* **2017**, *50*, 714–731. [CrossRef]
6. Zhang, Z.; Luo, S.; Yu, S.; Guan, Z.; Sun, R.; Wong, C.P. Significantly enhanced dielectric and energy storage performance of blend polymer-based composites containing inorganic 3D–network. *Mater. Des.* **2018**, *142*, 106–113. [CrossRef]
7. Moon, J.; Kim, B.; Choi, J.; Cho, M. Multiscale Study of the Relationship between Photoisomerization and Mechanical Behavior of Azo-Polymer Based on the Coarse-Grained Molecular Dynamics Simulation. *Macromolecules* **2019**, *52*, 2033–2049. [CrossRef]
8. Yang, S.; Kwon, S.; Lee, M.Y.; Cho, M. Molecular dynamics and micromechanics study of hygroelastic behavior in graphene oxide-epoxy nanocomposites. *Compos. Part B Eng.* **2019**, *164*, 425–436. [CrossRef]
9. Li, M.; Zhou, H.; Zhang, Y.; Liao, Y.; Zhou, H. Effect of defects on thermal conductivity of graphene/epoxy nanocomposites. *Carbon* **2018**, *130*, 295–303. [CrossRef]
10. Guan, L.; Zhao, L.; Wan, Y.J.; Tang, L.C. Three-dimensional graphene-based polymer nanocomposites: Preparation, properties and applications. *Nanoscale* **2018**, *10*, 14788–14811. [CrossRef]
11. Kausar, A.; Rafique, I.; Muhammad, B. Review of Applications of Polymer/Carbon Nanotubes and Epoxy/CNT Composites. *Polym.-Plast. Technol. Eng.* **2016**, *55*, 1167–1191. [CrossRef]
12. Park, H.; Choi, J.; Kim, B.; Yang, S.; Shin, H.; Cho, M. Toward the constitutive modeling of epoxy matrix: Temperature-accelerated quasi-static molecular simulations consistent with the experimental test. *Compos. Part B Eng.* **2018**, *142*, 131–141. [CrossRef]
13. Shin, H.; Choi, J.; Cho, M. An efficient multiscale homogenization modeling approach to describe hyperelastic behavior of polymer nanocomposites. *Compos. Sci. Technol.* **2019**, *175*, 128–134. [CrossRef]
14. Bao, S.; Tjong, S. Mechanical behaviors of polypropylene/carbon nanotube nanocomposites: The effects of loading rate and temperature. *Mater. Sci. Eng. A* **2008**, *485*, 508–516. [CrossRef]
15. Alian, A.R.; Meguid, S. Large-scale atomistic simulations of CNT-reinforced thermoplastic polymers. *Compos. Struct.* **2018**, *191*, 221–230. [CrossRef]
16. Alian, A.; El-Borgi, S.; Meguid, S. Multiscale modeling of the effect of waviness and agglomeration of CNTs on the elastic properties of nanocomposites. *Comput. Mater. Sci.* **2016**, *117*, 195–204. [CrossRef]
17. Zhang, W.; Deng, X.; Sui, G.; Yang, X. Improving interfacial and mechanical properties of carbon nanotube-sized carbon fiber/epoxy composites. *Carbon* **2019**, *145*, 629–639. [CrossRef]
18. Zeng, Q.; Yu, A.; Lu, G. Multiscale modeling and simulation of polymer nanocomposites. *Prog. Polym. Sci.* **2008**, *33*, 191–269. [CrossRef]

19. Paran, S.M.R.; Abdorahimi, M.; Shekarabi, A.; Khonakdar, H.A.; Jafari, S.H.; Saeb, M.R. Modeling and analysis of nonlinear elastoplastic behavior of compatibilized polyolefin/polyester/clay nanocomposites with emphasis on interfacial interaction exploration. *Compos. Sci. Technol.* **2018**, *154*, 92–103. [CrossRef]
20. Zhang, X.; Nguyen, H.; Daly, M.; Nguyen, S.T.; Espinosa, H.D. Nanoscale toughening of ultrathin graphene oxide-polymer composites: Mechanochemical insights into hydrogen-bonding/van der Waals interactions, polymer chain alignment, and steric parameters. *Nanoscale* **2019**, *11*, 12305–12316. [CrossRef]
21. Arash, B.; Park, H.S.; Rabczuk, T. Mechanical properties of carbon nanotube reinforced polymer nanocomposites: A coarse-grained model. *Compos. Part B Eng.* **2015**, *80*, 92–100. [CrossRef]
22. Xia, W.; Hansoge, N.K.; Xu, W.S.; Phelan, F.R.; Keten, S.; Douglas, J.F. Energy renormalization for coarse-graining polymers having different segmental structures. *Sci. Adv.* **2019**, *5*, eaav4683. [CrossRef] [PubMed]
23. Duan, K.; He, Y.; Li, Y.; Liu, J.; Zhang, J.; Hu, Y.; Lin, R.; Wang, X.; Deng, W.; Li, L. Machine-learning assisted coarse-grained model for epoxies over wide ranges of temperatures and cross-linking degrees. *Mater. Des.* **2019**, *183*, 108130. [CrossRef]
24. Noid, W.G. Perspective: Coarse-grained models for biomolecular systems. *J. Chem. Phys.* **2013**, *139*, 090901. [CrossRef]
25. Dirk, R.; Mathias, P.; Florian, M.P. Deriving effective mesoscale potentials from atomistic simulations. *J. Comput. Chem.* **2003**, *24*, 1624–1636.
26. Izvekov, S.; Roth, G.A. A multiscale coarse-graining method for biomolecular systems. *J. Phys. Chem. B* **2005**, *109*, 2469–2473. [CrossRef]
27. Noid, W.G.; Chu, J.W.; Ayton, G.S.; Krishna, V.; Izvekov, S.; Voth, G.A.; Das, A.; Andersen, H.C. The multiscale coarse-graining method. I. A rigorous bridge between atomistic and coarse-grained models. *J. Chem. Phys.* **2008**, *128*, 244114. [CrossRef]
28. Lyubartsev, A.P.; Laaksonen, A. Calculation of effective interaction potentials from radial distribution functions: A reverse Monte Carlo approach. *Phys. Rev. E* **2008**, *52*, 3730. [CrossRef]
29. Buehler, J.M. Mesoscale modeling of mechanics of carbon nanotubes: Self-assembly, self-folding, and fracture. *J. Mater. Res.* **2006**, *21*, 2855–2869. [CrossRef]
30. Li, Y.; Abberton, B.C.; Kroger, M.; Liu, W.K. Challenges in Multiscale Modeling of Polymer Dynamics. *Polymers* **2013**, *5*, 751–832. [CrossRef]
31. Zhao, J.; Jiang, J.W.; Wang, L.; Guo, W.; Rabczuk, T. Coarse-grained potentials of single-walled carbon nanotubes. *J. Mech. Phys. Solids* **2014**, *71*, 197–218. [CrossRef]
32. Arash, B.; Park, H.S.; Rabczuk, T. Coarse-grained model of the J-integral of carbon nanotube reinforced polymer composites. *Carbon* **2016**, *96*, 1084–1092. [CrossRef]
33. Mayo, S.L.; Olafson, B.D.; Goddard, W.A. DREIDING: A generic force field for molecular simulations. *J. Phys. Chem.* **1990**, *94*, 8897–8909. [CrossRef]
34. Dauber-Osguthorpe, P.; Roberts, V.A.; Osguthorpe, D.J.; Wolff, J.; Genest, M.; Hagler, A.T. Structure and energetics of ligand binding to proteins: Escherichia coli dihydrofolate reductase-trimethoprim, a drug-receptor system. *Proteins* **1988**, *4*, 31–47. [CrossRef] [PubMed]
35. Sun, H.; Mumby, S.J.; Maple, J.R.; Hagler, A.T. An ab initio CFF93 all-atom force field for polycarbonates. *J. Am. Chem. Soc.* **1994**, *116*, 2978–2987. [CrossRef]
36. Rappe, A.K.; Goddard, W.A., III. Charge equilibration for molecular dynamics simulations. *J. Phys. Chem.* **1991**, *95*, 3358–3363. [CrossRef]
37. Duan, K.; Zhang, J.; Li, L.; Hu, Y.; Zhu, W.; Wang, X. Diamond nanothreads as novel nanofillers for cross-linked epoxy nanocomposites. *Compos. Sci. Technol.* **2019**, *174*, 84–93. [CrossRef]
38. Li, Y.; Kröger, M. Viscoelasticity of carbon nanotube buckypaper: Zipping–unzipping mechanism and entanglement effects. *Soft Matter* **2012**, *8*, 7822–7830. [CrossRef]
39. Liu, X.; Lu, W.; Ayala, O.M.; Wang, L.P.; Karlsson, A.M.; Yang, Q.; Chou, T.W. Microstructural evolution of carbon nanotube fibers: Deformation and strength mechanism. *Nanoscale* **2013**, *5*, 2002–2008. [CrossRef]
40. Shen, Z.; Röding, M.; Kröger, M.; Li, Y. Carbon nanotube length governs the viscoelasticity and permeability of buckypaper. *Polymers* **2017**, *9*, 115. [CrossRef]
41. Sengab, A.; Picu, R.C. Mechanical behavior of carbon nanotube yarns with stochastic microstructure obtained by stretching buckypaper. *Compos. Sci. Technol.* **2018**, *166*, 54–65. [CrossRef]

42. Plimpton, S. Fast parallel algorithms for short-range molecular dynamics. *J. Comput. Phys.* **1995**, *117*, 1–19. [CrossRef]
43. Stukowski, A. Visualization and analysis of atomistic simulation data with OVITO–the Open Visualization Tool. *Model. Simul. Mater. Sci. Eng.* **2009**, *18*, 015012. [CrossRef]
44. Choi, J.; Shin, H.; Cho, M. A multiscale mechanical model for the effective interphase of SWNT/epoxy nanocomposite. *Polymer* **2016**, *89*, 159–171. [CrossRef]
45. Awasthi, A.P.; Lagoudas, D.C.; Hammerand, D.C. Modeling of graphene–polymer interfacial mechanical behavior using molecular dynamics. *Model. Simul. Mater. Sci. Eng.* **2008**, *17*, 015002. [CrossRef]
46. Li, Y.; Seidel, G.D. Multiscale modeling of the effects of nanoscale load transfer on the effective elastic properties of unfunctionalized carbon nanotube–polyethylene nanocomposites. *Model. Simul. Mater. Sci. Eng.* **2014**, *22*, 025023. [CrossRef]
47. Zhang, Y.; Zhuang, X.; Muthu, J.; Mabrouki, T.; Fontaine, M.; Gong, Y.; Rabczuk, T. Load transfer of graphene/carbon nanotube/polyethylene hybrid nanocomposite by molecular dynamics simulation. *Compos. Part B Eng.* **2014**, *63*, 27–33. [CrossRef]
48. Wang, T.; Dalton, A.B.; Keddie, J.L. Importance of Molecular Friction in a Soft Polymer-Nanotube Nanocomposite. *Macromolecules* **2008**, *41*, 7656–7661. [CrossRef]
49. Lordi, V.; Yao, N. Molecular mechanics of binding in carbon-nanotube–polymer composites. *J. Mater. Res.* **2000**, *15*, 2770–2779. [CrossRef]
50. Lebold, K.M.; Noid, W.G. Dual approach for effective potentials that accurately model structure and energies. *J. Chem. Phys.* **2019**, *150*, 234107. [CrossRef]
51. Lebold, K.M.; Noid, W.G. Systematic study of temperature and density variations in effective potentials for coarse-grained models of molecular liquids. *J. Chem. Phys.* **2019**, *150*, 014104. [CrossRef] [PubMed]
52. DeLyser, M.R.; Noid, W.G. Extending pressure-matching to inhomogeneous systems via local-density potentials. *J. Chem. Phys.* **2017**, *147*, 134111. [CrossRef] [PubMed]
53. Dwaikat, M.; Spitas, C.; Spitas, V. Predicting nonlinear stress-strain curves of unidirectional fibrous composites in consideration of stick-slip. *Compos. Part B Eng.* **2013**, *44*, 501–507. [CrossRef]
54. Shokrieh, M.M.; Rafiee, R. Investigation of nanotube length effect on the reinforcement efficiency in carbon nanotube based composites. *Compos. Struct.* **2010**, *92*, 2415–2420. [CrossRef]
55. Karimi, M.; Montazeri, A.; Ghajar, R. On the elasto-plastic behavior of CNT-polymer nanocomposites. *Compos. Struct.* **2017**, *160*, 782–791. [CrossRef]
56. Vu-Bac, N.; Rafiee, R.; Zhuang, X.; Lahmer, T.; Rabczuk, T. Uncertainty quantification for multiscale modeling of polymer nanocomposites with correlated parameters. *Compos. Part B Eng.* **2015**, *68*, 446–464. [CrossRef]
57. Gao, X.L.; Li, K. A shear-lag model for carbon nanotube-reinforced polymer composites. *Int. J. Solids Struct.* **2005**, *42*, 1649–1667. [CrossRef]
58. Li, X.; Gao, H.; Scrivens, W.A.; Fei, D.; Xu, X.; Sutton, M.A.; Reynolds, A.P.; Myrick, M.L. Nanomechanical characterization of single-walled carbon nanotube reinforced epoxy composites. *Nanotechnology* **2004**, *15*, 1416–1423. [CrossRef]

© 2019 by the authors. Licensee MDPI, Basel, Switzerland. This article is an open access article distributed under the terms and conditions of the Creative Commons Attribution (CC BY) license (http://creativecommons.org/licenses/by/4.0/).

nanomaterials

MDPI

Article

Molecular Dynamics Simulation on Mechanical and Piezoelectric Properties of Boron Nitride Honeycomb Structures

Lu Xie [1,*], Tianhua Wang [1], Chenwei He [2], Zhihui Sun [1,*] and Qing Peng [3,*]

1 School of Mechanical Engineering, University of Science and Technology Beijing, Beijing 100083, China
2 Reactor Engineering and Safety Research Center, China Nuclear Power Technology Research Institute Co., Ltd., Shenzhen 518031, China
3 Nuclear Engineering and Radiological Sciences, University of Michigan, Ann Arbor, MI 48108, USA
* Correspondence: xielu@ustb.edu.cn (L.X.); sunzhihui@ustb.edu.cn (Z.S.); peng@umich.edu (Q.P.)

Received: 29 June 2019; Accepted: 18 July 2019; Published: 21 July 2019

Abstract: Boron nitride honeycomb structure is a new three-dimensional material similar to carbon honeycomb, which has attracted a great deal of attention due to its special structure and properties. In this paper, the tensile mechanical properties of boron nitride honeycomb structures in the zigzag, armchair and axial directions are studied at room temperature by using molecular dynamics simulations. Effects of temperature and strain rate on mechanical properties are also discussed. According to the observed tensile mechanical properties, the piezoelectric effect in the zigzag direction was analyzed for boron nitride honeycomb structures. The obtained results showed that the failure strains of boron nitride honeycomb structures under tensile loading were up to 0.83, 0.78 and 0.55 in the armchair, zigzag and axial directions, respectively, at room temperature. These findings indicated that boron nitride honeycomb structures have excellent ductility at room temperature. Moreover, temperature had a significant effect on the mechanical and tensile mechanical properties of boron nitride honeycomb structures, which can be improved by lowering the temperature within a certain range. In addition, strain rate affected the maximum tensile strength and failure strain of boron nitride honeycomb structures. Furthermore, due to the unique polarization of boron nitride honeycomb structures, they possessed an excellent piezoelectric effect. The piezoelectric coefficient e obtained from molecular dynamics was 0.702 C/m^2, which was lower than that of the monolayer boron nitride honeycomb structures, $e = 0.79\ C/m^2$. Such excellent piezoelectric properties and failure strain detected in boron nitride honeycomb structures suggest a broad prospect for the application of these new materials in novel nanodevices with ultrahigh tensile mechanical properties and ultralight-weight materials.

Keywords: boron nitride honeycomb; molecular dynamics simulation; mechanical property; piezoelectric property

1. Introduction

Boron nitride (BN) has a similar structure to graphene and exhibits excellent mechanical and electrical properties [1–5]. The two-dimensional BN films have been successfully stripped out by using micromechanical cleavage. These structures reveal high crystal quality and macroscopic continuity [6]. BN nanobelts are fabricated by a simple ZnS nanoribbon templating method and possess good optical properties [7]. The wide application of two-dimensional materials in various fields has attracted the interest of different research groups in three-dimensional materials. A novel boron nitride honeycomb (BNHC) structure consisting of zigzag-edged BN nanosheets is proposed by Wu et al. [8] and they confirmed structural stability of this material. In particular, carbon honeycomb (CHC) structures which are similar to BNHCs, have been successfully fabricated [9]. These honeycomb structures can be used not only for storing various gases and liquids but also as a matrix for new composite materials.

Since the first report by Wang et al. on a prototype nanogenerator based on zinc oxide nanowires [10], piezoelectric nanomaterials have received extensive attention [11–13]. It has been found that BN belongs to the piezoelectric materials [14] and shows good piezoelectric effect [4]. The piezoelectric effect means that when an external pressure is applied to a piezoelectric material, a potential difference is generated on the surface of the material. Conversely, a piezoelectric material deforms when an external electric field is applied to it. The essence is that when pressure is exploited to a piezoelectric material, the non-centrosymmetric ions inside the crystal starts to be polarized and results in a potential difference. Owing to the simultaneous possession of piezoelectricity and semiconductor properties, the piezopotential created in the crystal has a strong effect on the carrier transport at the interface/junction. The piezoelectric potential produced by the mechanical deformation of a piezoelectric material can be used as the gate voltage to change the carrier transmission characteristics [15] and thereby improves the performance of photovoltaic devices such as nanosensors [16], nanogenerators [17], nanotransistors [18] and so forth. According to the density functional theory (DFT) calculations, it is found that BN nanosheets exhibit stronger piezoelectric coupling than conventional bulky Wurtzite structures [4]. The piezoelectric effect of BNHC structures has been analyzed by using a combination of finite element and molecular dynamics simulations and studies have indicated that BNHC structures provide a good piezoelectric effect and piezopotential properties which can be efficiently adjusted by regulating the lattice constant [19].

Since mechanical properties of a material directly affect its application in various fields, therefore, it is necessary to study this important parameter. There are many studies on the mechanical properties of BN nanotubes, for example, the elastic properties of an individual multi-wall BN nanotube is determined experimentally and the results confirm that these nanotubes are highly crystalline with few defects [20]. Moreover, mechanical properties of monolayer systems of honeycomb structures are investigated by using an equation of state (EOS). The results indicate that graphene is the most elastic, followed by BN films and both materials have considerable strength [21]. Inspired by the excellent mechanical properties and wide applications of 2D materials, such as BN nanosheets and graphene, it is reasonable to build three-dimensional materials with excellent mechanical properties. Some studies have shown that mechanical properties of CHC structures bear a strong cell-size effect and anisotropy. The failure stress and failure strain in different directions decrease with enhancing the cell-size [22]. However, for BNHC structures, scholars focus on their electrical properties instead of mechanical properties. The mechanical properties of BNHC structures are the premise to ensure other properties. Therefore, mechanical properties of BNHC structures require further attentions. There are also relatively few analyses on the mechanical properties of BNHC structures using molecular dynamics (MD) simulations.

Motivated by these ideas, a boron nitride honeycomb structure was constructed based on the BN nanosheets in this paper. The structures possessed good mechanical properties and piezoelectric effect and tensile mechanical properties of the structures in different directions were studied by using MD simulations. The effects of temperature and strain rate on tensile mechanical properties were also analyzed. Based on the above studies, the piezoelectric properties of BNHCs with a specific lattice constant in the zigzag direction were further affirmed.

2. Materials and Methods

In this research, a new three-dimensional boron nitride honeycomb structure is established as shown in Figure 1. It can be viewed as consisting of zigzag-edged BN nanoribbons with sp^2 bonding in the wall and sp^3 bonding in junction. The cross-section perpendicular to the axial (cell axis) is a honeycomb structure based on regular hexagon with a side length 5.8 Å.

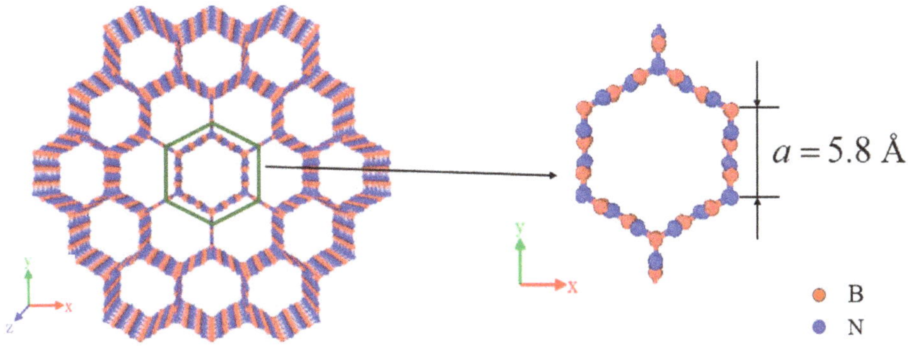

Figure 1. The graphical structure of boron nitride honeycomb.

Based on the established model, the tensile mechanical properties and piezoelectric behaviors were studied by MD simulations. Herein, all MD simulations were conducted using the large-scale atomic/molecular massively parallel simulator (LAMMPS) [23]. The structural analysis and data processing of BNHCs were conducted by using the visualization software OVITO [24]. In all the simulations, the interactions between B and N atoms were described by Tersoff potential [25]. In the present simulations, the values of parameters in the Tersoff potential were taken from Reference [26]. This Tersoff potential has been successfully employed to evaluate the mechanical and piezoelectric properties of BN [27,28]. We tested three potentials of Tersoff, extep and Reaxff for BNHCs and found that the Tersoff potential is the most suitable (see the Supplementary Files for details). The curves of the total energy and temperature of BNHCs were obtained (see Figure S1 in the Supplementary Files). Time step is 0.5 fs. The isothermal-isobaric (NPT) ensemble was utilized to update the positions and velocities of the atoms by using the Nosé-Hoover temperature thermostat [29]. In addition, the velocity Verlet algorithm was employed to integrate the Hamiltonian equations of the determined motion. Before analysis of performance of the model, the conjugate gradient algorithm was used to calculate the energy minimization of BNHCs to obtain their stable states. Then, BNHCs were relaxed at room temperature (300 K) for 25 ps to obtain a state of equilibrium. After relaxation, the model was contained to a total of 2016 atoms and size of the model was 30.32 Å × 35.01 Å × 29.52 Å.

2.1. MD Simulations for Tensile Mechanical Properties of BNHCs

First, the equilibrium system was gradually heated from 300 to 5000 K at a heating rate of 4.7×10^{12} K·s^{-1} and the melting characteristics of BNHCs were analyzed. After the system was fully relaxed, the mechanical properties of BNHCs were analyzed. In this paper, the tensile mechanical properties of BNHCs in the zigzag (x-axis), armchair (y-axis) and axial (z-axis) directions were studied. Thereafter, the influences of temperature and strain rate were analyzed on the mechanical properties of BNHCs. The strain load was applied in different directions and the Young's modulus and the maximum tensile strength were obtained according to the stress-strain curves. The engineering strain rate was taken as 10^9 s^{-1} in the simulation.

2.2. MD Simulations for Piezoelectric Properties of BNHCs

According to the studies of mechanical properties, the piezoelectric behaviors of the zigzag direction of BNHCs were investigated. Piezoelectric effect refers to the influence of external pressure on a piezoelectric material, in which the material produces a potential difference on its surface and causes deformation of the material due to the generation of an external electric field. Therefore, the piezoelectric effect is essentially dependent on the deformation of the material due to external forces, which polarizes the internal material. The piezoelectric constant is a parameter that characterizes the polarization of a piezoelectric material under an external pressure. Herein, the MD simulations were used to apply the strain load to the zigzag direction of BNHCs. The piezoelectric constant was obtained by calculating the polarization inside BNHCs and, then, piezoelectric characteristics were evaluated.

At a temperature of 300 K, a strain load was applied to the zigzag direction of BNHCs and the polarization of P and strain of ε were recorded in the zigzag direction. The piezoelectric constant of e was obtained by calculating the slopes of the polarization (P) and strain (ε) [30].

3. Results and Discussion

Before analysis of the mechanical properties, the system was warmed-up. The curve of the total energy, volume and temperature of BNHCs were obtained at a heating rate of 4.7×10^{12} K·s^{-1}, as shown in Figure 2.

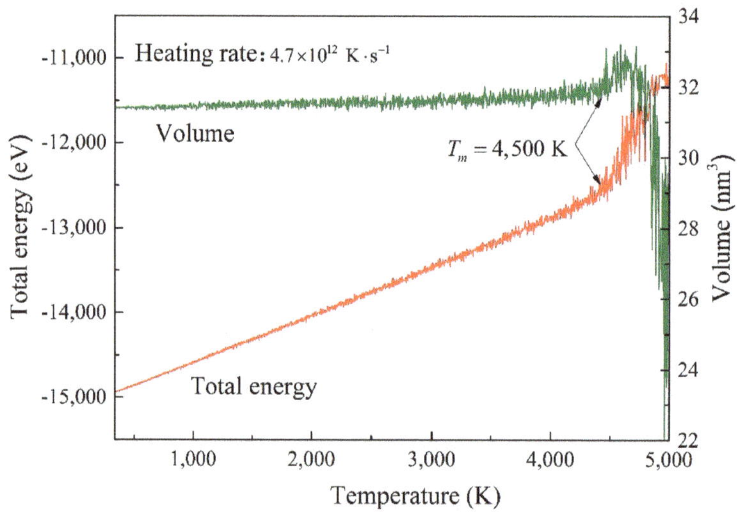

Figure 2. The curves of total energy and volume versus temperature during heating of boron nitride honeycombs (BNHCs).

Figure 2 shows that the total energy and volume of BNHCs increased linearly with increasing temperature during the heating process. However, when the temperature reached about 4500 K, the total energy inclined significantly, and the maximum volume was also reached. When the temperature continued to increase, the system-energy was also enhanced and the volume dropped sharply. This finding indicated that the phase of BNHC structure was changed at about 4500 K and it began to melt. The obtained melting point of BNHCs by MD simulations was around 4500 K. It is found that CHCs maintains structural stability at 2000 K [31], which confirmed the high thermal stability of BNHCs.

3.1. Tensile Mechanical Properties of BNHCs

In this section, the mechanical properties of BNHCs under tensile loading along the zigzag (x-axis), armchair (y-axis) and axial (z-axis) directions were studied at room temperature (300 K). At a strain rate of 10^9 s^{-1}, the stress-strain curve along the zigzag direction is shown in Figure 3.

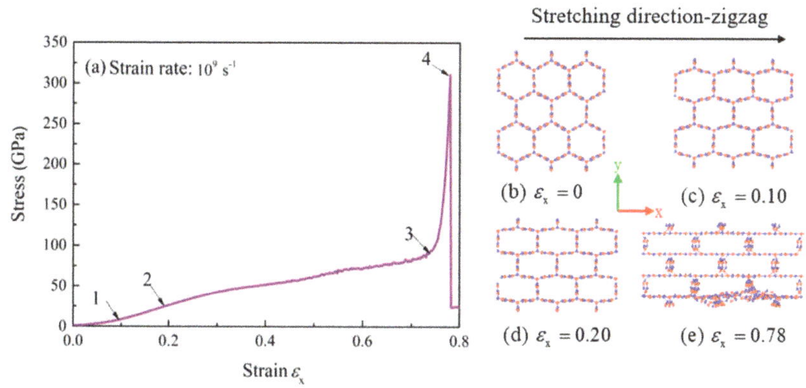

Figure 3. (**a**) The stress-strain curve of BNHCs under tensile loading along the zigzag direction; (**b**)–(**e**) Deformation of BNHCs under different tensile strains along the zigzag direction.

It can be seen from the stress-strain curve in Figure 3a that the change of the bond angle played a leading role in the initial stage of stretching. At the stretching of 0.1 (point 1), the bond angle and B-N bond length together caused the fast increase of stress. When the stretching was continued until the strain of 0.2 (point 2), the rate of stress was slowed down, because the increase of strain had been due to the elongation of B-N bond. The tensile strength of BNHCs in the zigzag direction did not show a good yielding stage and the strengthening phase began from the strain of 0.76 (point 3). When the strain was increased to 0.78 (point 4), the stress was lowered sharply, and the material failed. The Young's modulus of BNHCs under tensile loading along the zigzag direction was calculated as about 146 GPa and the maximum tensile strength of 310 GPa was observed at the strain of 0.78.

Figure 3b–e show the tensile deformation process of BNHCs in the zigzag direction. As these figures show, BNHCs had a regular hexagonal structure when unstretched and this structure was basically maintained before the strain of 0.1. This finding revealed that the stress was increased linearly with strain; when the strain reached 0.2, the cross section of BNHCs was transformed into a rectangular structure until failure occurred at the strain of 0.78. This perception structurally demonstrated that elongation of the bond length caused an increase in the strain during the above stretching phase.

The tensile stress-strain curve of BNHCs along the armchair is shown in Figure 4a. There are four stages in the stretching process: At the beginning of stretching, the stress was increased almost linearly with strain and increasing of strain during this process was mainly due to the change of B-N bond angle. When the strain reached 0.1 (point 1), the rate of stress-enhancement was increased because the bond length and bond angle are changing. When the strain approached 0.23 (point 2), the maximum rate of stress was attained, mainly due to the significant changes of B-N bond length. When the strain continued to increase up to 0.5 (point 3), only the bond length was altered, and the rate of stress was decreased. The maximum tensile strength of 156 GPa was achieved as the strain reached 0.83 (point 4). The strain increases and the structure eventually failed. The Young's modulus of about 173 GPa was calculated for BNHCs under tensile loading along the armchair direction.

Figure 4b–e describe the deformation process of BNHCs as it is stretched along the armchair direction. It can be seen that when the stretching to strain was 0.1, the regular hexagonal cross section of BNHCs started to change due to alteration of the bond angle. When the strain was 0.23, the BNHCs completely deviated from the regular hexagon. Eventually, the structure was destroyed, and the material failed when the strain increased to 0.83.

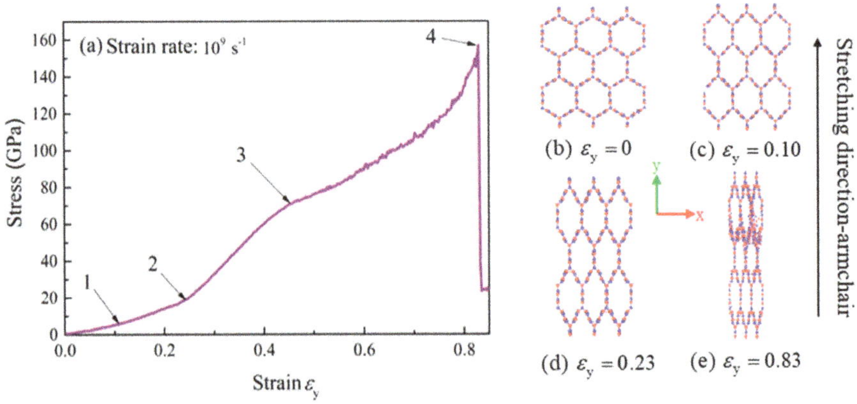

Figure 4. (**a**) The stress-strain curve of BNHCs under tensile loading along the armchair direction; (**b–e**) Deformation of BNHCs under different tensile strains along the armchair direction.

The stress-strain curve of BNHCs under tensile loading along the axial direction is shown in Figure 5a. The four stages of the stretching process were included (1) approximately linear increase of the stress with strain at the beginning of stretching, similar to the linear phase. (2) The strain showed a small change with strain at the stage of 0.28–0.49, which was approximated as the yield stage. The stresses were about 57 and 74 GPa at points 1 and 2, respectively. (3) When the strain was in the stage of 0.49–0.55, the deformation of the material was enhanced as the strain was increased, the strengthening stage appeared and the maximum tensile strength of 247 GPa was attained at point 3. (4) Finally, the structure was failed by continuation of stretch. The Young's modulus of BNHCs under tensile loading along the axial direction was calculated as about 334 GPa.

Figure 5b–e describes the tensile deformation process for BNHCs along the axial direction. It was found that the axial direction grew in size as the strain increased, until a strain of 0.55, in which the structure was destroyed, and the material failed.

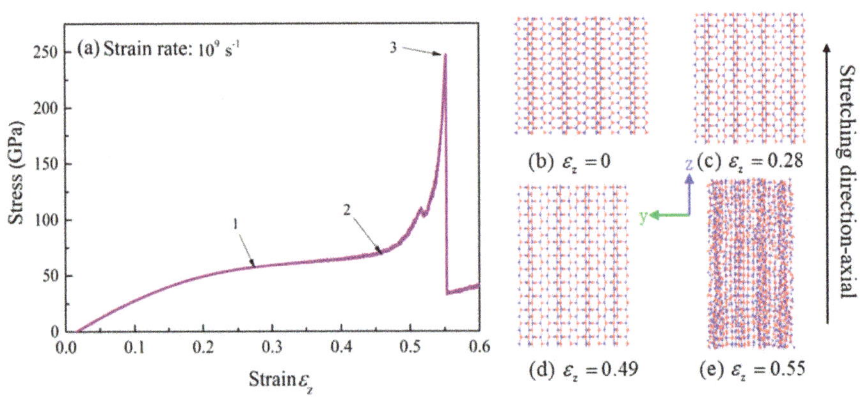

Figure 5. (**a**) The stress-strain curve of BNHCs under tensile loading along the axial direction; (**b–e**) Deformation of BNHCs under different tensile strains along the axial direction.

Table 1 lists the Young's modulus, failure strain and maximum tensile strength of BNHCs under tensile in three different directions at room temperature. It can be envisaged that the largest Young's modulus was obtained in the axial direction, whereas no significantly different values were gained for the zigzag and armchair directions. The axial direction showed the smallest failure strain, indicating that ductility in the axial direction was not good and ductility of the armchair direction was slightly higher than that of the zigzag direction. The maximum tensile strength was observed in the zigzag direction. The tensile processes and the evolution of shear stains of BNHCs in the zigzag, armchair and cell axis directions are presented, respectively (see Video S1–S3 in the Supplementary Files).

Table 1. Mechanical proprieties of BNHCs under the tensile loading along different directions.

Stretching Direction	Young's Modulus (GPa)	Failure Strain	Maximum Tensile Strength (GPa)
zigzag-x	146	0.78	310
armchair-y	173	0.83	156
axial-z	334	0.55	247

The failure strain of 0.76–0.81 was calculated for BNHCs along the armchair direction by using the combination of finite element method and molecular dynamics simulations [19]. The results of this investigation were consistent with the previous reports. The failure strains of CHC in the zigzag, armchair and cell axis directions were 0.204, 0.320 and 0.225, respectively. Moreover, the maximum tensile strengths of CHC in the zigzag, armchair and cell axis directions were 23.7, 22.4 and 55.3 GPa, respectively [32]. The failure strain of BNHCs reached about 0.8, which was significantly higher than that of CHC; the maximum tensile strength was also significantly higher than that of CHC, Compared to CHC, BNHCs showed high failure strain and improved mechanical properties. In addition, the failure strain of BNHCs was significantly higher than that of other piezoelectric materials. The failure strains of MoS_2 nanosheets have been about 0.09 and 0.12 in the zigzag and armchair directions [33], whereas the failure strains of about 0.2 and 0.15 are obtained for GaN [34] and ZnO [35] nanowires, respectively.

3.2. *Effect of Temperature on the Tensile Mechanical Properties of BNHCs*

In practical engineering, the mechanical properties of materials often determine their applications. Many factors such as composition, stress state, nature of the load, environmental conditions and temperature affect properties of the material. In this section, effects of temperature and strain rate on the mechanical properties of BNHCs are mainly studied by using MD simulations.

Figure 6 shows the stress-strain curves of BNHCs under tensile loading in three different directions over a temperature range of 100–700 K. According to Figure 6a, when the structure was stretched in the zigzag direction, the maximum tensile strength of 310 GPa was attained at room temperature (300 K); but, this parameter was reduced 31.6% and reached to 98 GPa as the temperature raised-up to 700 K. Accordingly, the maximum tensile strength of 249 GPa was achieved at 100 K, which is 80.3% of that at room temperature. From the failure strain points of view, the failure strains of 0.775, 0.78, 0.772 and 0.73 were obtained at 100, 300, 500 and 700 K, respectively. Based on the above results, the maximum tensile strength and failure strain of BNHCs were achieved in the zigzag direction at room temperature, indicating the best strength at room temperature. However, the maximum tensile strength and failure strain were higher at low temperature. Therefore, the best properties of BNHCs in the zigzag direction were observed at low temperatures in the range of 100–700 K and the best tensile property was attained at room temperature.

It can be seen from Figure 6b that a good temperature dependence was gained when the BNHCs was stretched along the armchair direction. The maximum tensile strength and failure strain were decreased with enhancing temperature. The maximum tensile strength ranged from 120 to 184 GPa and the failure strain changed from 0.79 to 0.87, indicating that the excellent ductility of BNHCs was perceived in the armchair direction. The maximum tensile strength did not fluctuate significantly at different temperatures and the failure strain was greater than that in the zigzag direction.

As Figure 6c shows, the maximum tensile strength and failure strain in the cell-axis direction were lowered as the temperature grown up during stretching along the axial direction. The maximum range of tensile strength was78–341 GPa and the failure strain was 0.46–0.55. As the temperature was increased, the maximum tensile strength fluctuated greatly. The maximum tensile strength at 700 K was decreased 31.5% at room temperature. However, when the temperature was 100 K, the maximum tensile strength was 1.38 times of that at room temperature. This finding showed that similar to the zigzag direction, the strength was significantly higher at low temperatures.

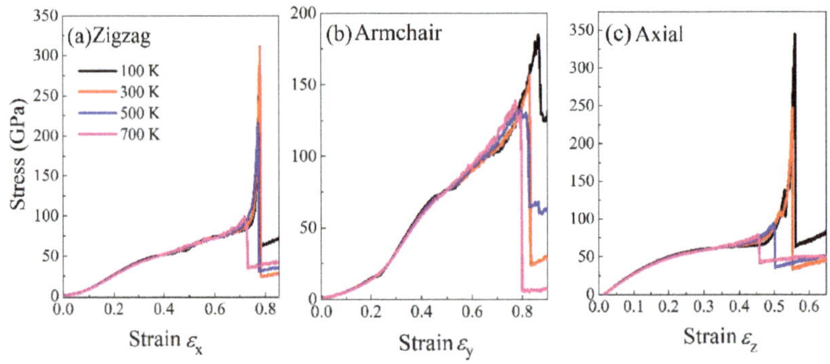

Figure 6. The stress-strain curves of BNHCs under tensile loading at various temperatures along (**a**) zigzag, (**b**) armchair and (**c**) axial directions, respectively.

3.3. Effect of Strain Rate on the Tensile Mechanical Properties of BNHCs

To some extent, the strain rate also affects the mechanical properties of the material. Figure 7 show tensile stress-strain curves at three different strain rates. The strain rate slightly affected the tensile properties but did not show an obvious regularity. The stress-strain curve was unchanged in the initial stage of stretching and the strain rate only influenced the maximum tensile strength and failure strain. In the zigzag direction (Figure 7a), the maximum tensile strength was first decreased and then grew up as the strain rate was enhanced. In the armchair direction, as the strain rate was developed, the maximum tensile strength and failure strain rate were increased. In the axial direction, it reached to 247 GPa at a strain rate of 10^9 s^{-1} and as the strain rate was inclined, the maximum tensile strength was decreased to 115 GPa. In short, the strain rate mainly affected the maximum tensile strength and failure strain of BNHCs. The stress-strain curve almost coincided in the initial stage of stretching. Due to different strain rates, the material broken up at different moments and resulted in various maximum tensile strengths.

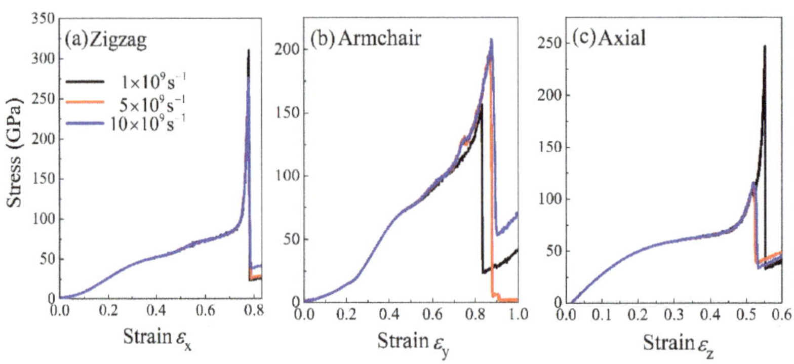

Figure 7. The stress-strain curves of BNHCs under tensile loading at various stain rates along (**a**) zigzag, (**b**) armchair and (**c**) axial directions, respectively.

3.4. Piezoelectric Properties of BNHCs

In order to study the piezoelectric properties of BNHCs, first the strain load was applied to the monolayer of BNHCs along the zigzag direction. It was found that center of the triangle composed of three B atomic charges inside the structure did not coincide with the center of the triangle composed of three N atomic charges and the electric dipole moment caused polarization of the overall structure. Figure 8 shows a diagram for the distribution of atomic charge structures when BN nanosheet is stretched.

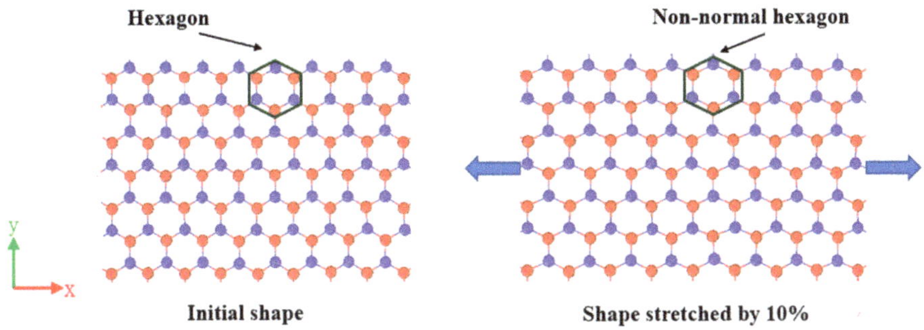

Figure 8. Atomic charge distribution of BN nanosheet under tensile loading.

The piezoelectric constant is a parameter that characterizes the internal polarization of a piezoelectric material and can be obtained by the slopes of polarization, P and strain of ε. Therefore, polarization at different strains was calculated and curves of the polarization and strain were obtained for BN nanosheet, as shown in Figure 9.

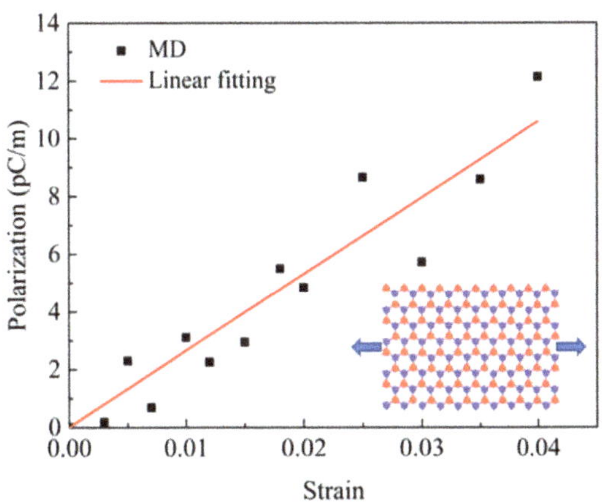

Figure 9. The polarization-strain curves of BN nanosheet under tensile loading along the zigzag direction obtained from molecular dynamics (MD) simulations.

From the above results, it can be deduced that the piezoelectric constant of BN nanosheet was $e = 2.65 \times 10^{-10}$ C/m, in the range of $1.38 \times 10^{-10} - 3.71 \times 10^{-10}$ C/m, which correlates well with the previous study [4]. The thickness of BN nanosheet was 0.33 nm during simulation and piezoelectric coefficient of the monolayer BNHCs was calculated as $e = 0.79$ C/m^2. Some researchers have provided the piezoelectric coefficient of $e = 1.92 \times 10^{-10}$ C/m for a BN nanosheet by molecular dynamics simulations. Since the thickness was 0.95Å, the piezoelectric constant after removing the thickness effect was attained as $e = 2.1$ C/m^2 [19].

Since BNHCs is composed of BN nanosheets, polarization of BNHCs, which is due to the piezoelectric effect, can be achieved as the summation of polarization of BN nanosheets. According to the above methods, the piezoelectric properties of multilayer BNHCs were analyzed and the curves of P and ε for BNHCs were obtained, as shown in Figure 10. The piezoelectric constant of the multilayer BNHCs was calculated as $e = 0.702\ \mathrm{C/m^2}$. Findings revealed that the piezoelectric coefficient of the multilayer BNHCs was slightly smaller than that of the monolayer counterpart. The classical piezoelectric theory [30] mentions that when an external pressure is applied to a piezoelectric material, its internal structure would be deformed, resulting in separation of the positive and negative ions inside the material and formation of a dipole. The opposite magnetic poles in the material cancel each other and charges may appear on the surface of the material. According to this theory, it be concluded that polarization between the inner of BNHCs canceled each other, which may cause the polarization phenomenon of the multilayer BNHCs to be smaller than that of the monolayer counterpart, resulting in a slightly smaller piezoelectric coefficient.

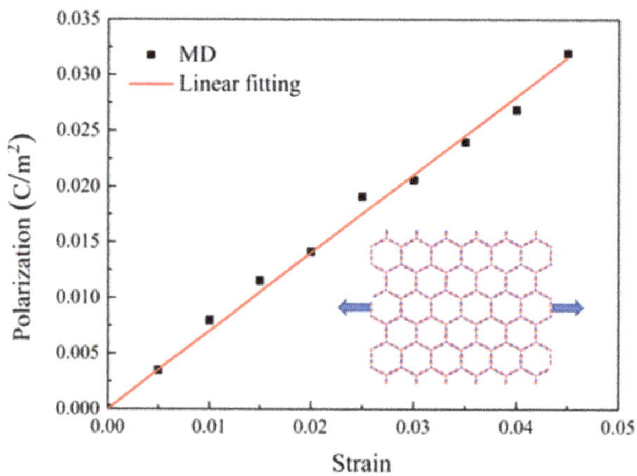

Figure 10. The polarization-strain curves of BNHCs under tensile loading along the zigzag direction obtained from MD simulations.

4. Conclusions

In this paper, the mechanical properties of BNHCs under tensile loading in three different directions of zigzag, armchair and axial directions were studied using molecular dynamics simulations. Effects of temperature and strain rate on the tensile mechanical properties were analyzed. The piezoelectric properties of BNHCs in the zigzag direction were also studied. Our results showed that BNHCs possesses excellent mechanical properties along with large failure strain. The failure strains of 0.83, 0.78 and 0.55 were attained in the armchair, zigzag and axial directions, respectively. The structure also showed a high tensile strength with maximum tensile strengths of 310, 156 and 247 GPa in the zigzag, armchair and axial directions, respectively and Young's modulus in the three mentioned directions were 146, 173 and 334 GPa, respectively. In the temperature range of 100–700 K, the mechanical properties of BNHCs at low temperature were higher than those at a high temperature. The maximum tensile strength and failure strain of BNHCs in the zigzag direction at room temperature were the highest, indicating the best strength at room temperature. In the other two directions, the maximum tensile strength and failure strain were decreased with enhancing temperature. At the initial stage of stretching, the strain rate had a little effect on the tensile properties. As the strain was increased to near the failure strain, the strain rate affected the maximum tensile strength and failure strain of BNHCs. In the armchair direction, as the strain rate was increased, the stress also inclined. Whereas, the stress was first decreased and then increased in the zigzag and cell-axis directions as the strain rate was grown up. The structure possessed an excellent piezoelectric effect due to its special internal polarization. The piezoelectric coefficient of the monolayer BNHCs was calculated as $e = 0.79\ \mathrm{C/m^2}$ and the piezoelectric coefficient of 0.702 $\mathrm{C/m^2}$ was attained for BNHCs with the thickness of 29.54 Å, which is slightly lower than that of the monolayer BNHCs.

The results of this study showed that BNHCs possessed excellent mechanical and piezoelectric properties and it may have great applications in new nanodevices with ultrahigh tensile mechanical properties and ultralight-weight materials.

Supplementary Materials: The following are available online at http://www.mdpi.com/2079-4991/9/7/1044/s1, Figure S1: The curves of total energy versus temperature during heating of BNHCs; Video S1: Tensile in zigzag direction; Video S2: Tensile in armchair direction; Video S3: Tensile in cell axis direction.

Author Contributions: Conceptualization, L.X. and Q.P.; methodology, C.H. and Z.S.; formal analysis, L.X.; investigation, T.W. and C.H.; writing—original draft preparation, T.W.; writing—review and editing, L.X. and Q.P.; visualization, C.H.; supervision, Q.P. and Z.S.; funding acquisition, L.X.

Funding: This research was funded by National Natural Science Foundation of China (21703007) and Fundamental Research Funds for the Central Universities (FRF-GF-18-011B).

Conflicts of Interest: The authors declare no conflict of interest.

References

1. Golberg, D.; Bando, Y.; Huang, Y.; Terao, T.; Mitome, M.; Tang, C.; Zhi, C. Boron Nitride Nanotubes and Nanosheets. *ACS Nano* **2010**, *4*, 2979–2993. [CrossRef] [PubMed]
2. Lin, Y.; Connell, J.W. Advances in 2D boron nitride nanostructures: Nanosheets, nanoribbons, nanomeshes, and hybrids with graphene. *Nanoscale* **2012**, *4*, 6908–6939. [CrossRef] [PubMed]
3. Pakdel, A.; Zhi, C.; Bando, Y.; Golberg, D. Low-dimensional boron nitride nanomaterials. *Mater. Today* **2012**, *15*, 256–265. [CrossRef]
4. Duerloo, K.-A.N.; Ong, M.T.; Reed, E.J. Intrinsic Piezoelectricity in Two-Dimensional Materials. *J. Phys. Chem. Lett.* **2012**, *3*, 2871–2876. [CrossRef]
5. Pakdel, A.; Bando, Y.; Golberg, D. Nano boron nitride flatland. *Chem. Soc. Rev.* **2014**, *43*, 934–959. [CrossRef] [PubMed]
6. Novoselov, K.S.; Jiang, D.; Schedin, F.; Booth, T.J.; Khotkevich, V.V.; Morozov, S.V.; Geim, A.K. Two-dimensional atomic crystals. *Proc. Natl. Acad. Sci. USA* **2005**, *102*, 10451–10453. [CrossRef] [PubMed]
7. Chen, Z.-G.; Zou, J.; Liu, G.; Li, F.; Wang, Y.; Wang, L.; Yuan, X.-L.; Sekiguchi, T.; Cheng, H.-M.; Lu, G.Q. Novel Boron Nitride Hollow Nanoribbons. *ACS Nano* **2008**, *2*, 2183–2191. [CrossRef]
8. Wu, M.; Wu, X.; Pei, Y.; Wang, Y.; Zeng, X.C. Three-dimensional network model of carbon containing only sp2-carbon bonds and boron nitride analogues. *Chem. Commun.* **2011**, *47*, 4406–4408. [CrossRef]
9. Krainyukova, N.V.; Zubarev, E.N. Carbon Honeycomb High Capacity Storage for Gaseous and Liquid Species. *Phys. Rev. Lett.* **2016**, *116*, 055501. [CrossRef]
10. Wang, Z.L.; Song, J. Piezoelectric Nanogenerators Based on Zinc Oxide Nanowire Arrays. *Science* **2006**, *312*, 242–246. [CrossRef]
11. Wu, W.; Wang, Z.L. Piezotronics and piezo-phototronics for adaptive electronics and optoelectronics. *Nat. Rev. Mater.* **2016**, *1*, 16031. [CrossRef]
12. Zhang, J.; Meguid, S.A. Piezoelectricity of 2D nanomaterials: Characterization, properties, and applications. *Semicond. Sci. Technol.* **2017**, *32*, 043006. [CrossRef]
13. Espinosa, H.D.; Bernal, R.A.; Minary-Jolandan, M. A Review of Mechanical and Electromechanical Properties of Piezoelectric Nanowires. *Adv. Mater.* **2012**, *24*, 4656–4675. [CrossRef] [PubMed]
14. Mele, E.J.; Král, P. Electric Polarization of Heteropolar Nanotubes as a Geometric Phase. *Phys. Rev. Lett.* **2002**, *88*, 056803. [CrossRef] [PubMed]
15. Wang, Z.L. Progress in Piezotronics and Piezo-Phototronics. *Adv. Mater.* **2012**, *24*, 4632–4646. [CrossRef] [PubMed]
16. Qi, J.; Lan, Y.-W.; Stieg, A.Z.; Chen, J.-H.; Zhong, Y.-L.; Li, L.-J.; Chen, C.-D.; Zhang, Y.; Wang, K.L. Piezoelectric effect in chemical vapour deposition-grown atomic-monolayer triangular molybdenum disulfide piezotronics. *Nat. Commun.* **2015**, *6*, 7430. [CrossRef] [PubMed]

17. Kim, S.K.; Bhatia, R.; Kim, T.-H.; Seol, D.; Kim, J.H.; Kim, H.; Seung, W.; Kim, Y.; Lee, Y.H.; Kim, S.-W. Directional dependent piezoelectric effect in CVD grown monolayer MoS2 for flexible piezoelectric nanogenerators. *Nano Energy* **2016**, *22*, 483–489. [CrossRef]
18. Gao, Z.; Zhou, J.; Gu, Y.; Fei, P.; Hao, Y.; Bao, G.; Wang, Z.L. Effects of piezoelectric potential on the transport characteristics of metal-ZnO nanowire-metal field effect transistor. *J. Appl. Phys.* **2009**, *105*, 113707. [CrossRef]
19. Zhang, J. Boron nitride honeycombs with superb and tunable piezopotential properties. *Nano Energy* **2017**, *41*, 460–468. [CrossRef]
20. Chopra, N.G.; Zettl, A. Measurement of the elastic modulus of a multi-wall boron nitride nanotube. *Solid State Commun.* **1998**, *105*, 297–300. [CrossRef]
21. Andrew, R.C.; Mapasha, R.E.; Ukpong, A.M.; Chetty, N. Mechanical properties of graphene and boronitrene. *Phys. Rev. B* **2012**, *85*, 125428. [CrossRef]
22. Pang, Z.; Gu, X.; Wei, Y.; Yang, R.; Dresselhaus, M.S. Bottom-up Design of Three-Dimensional Carbon-Honeycomb with Superb Specific Strength and High Thermal Conductivity. *Nano Lett.* **2017**, *17*, 179–185. [CrossRef]
23. Plimpton, S. Fast Parallel Algorithms for Short-Range Molecular Dynamics. *J. Comput. Phys.* **1995**, *117*, 1–19. [CrossRef]
24. Stukowski, A. Visualization and analysis of atomistic simulation data with OVITO–The Open Visualization Tool. *Model. Simul. Mater. Sci. Eng.* **2010**, *18*, 015012. [CrossRef]
25. Tersoff, J. Modeling solid-state chemistry: Interatomic potentials for multicomponent systems. *Phys. Rev. B* **1989**, *39*, 5566–5568. [CrossRef]
26. Liao, M.-L.; Wang, Y.-C.; Ju, S.-P.; Lien, T.-W.; Huang, L.-F. Deformation behaviors of an armchair boron-nitride nanotube under axial tensile strains. *J. Appl. Phys.* **2011**, *110*, 054310. [CrossRef]
27. Mortazavi, B.; Rémond, Y. Investigation of tensile response and thermal conductivity of boron-nitride nanosheets using molecular dynamics simulations. *Phys. E Low Dimens. Syst. Nanostruct.* **2012**, *44*, 1846–1852. [CrossRef]
28. Zhang, J.; Meguid, S.A. Effect of number of layers upon piezoelectric behaviour of multi-walled boron nitride nanotubes. *J. Phys. D Appl. Phys.* **2015**, *48*, 495301. [CrossRef]
29. Nosé, S. A unified formulation of the constant temperature molecular dynamics methods. *J. Chem. Phys.* **1984**, *81*, 511–519. [CrossRef]
30. Arnau, A.; Soares, D. Fundamentals of Piezoelectricity. In *Piezoelectric Transducers and Applications*; Springer: Berlin/Heidelberg, Germany, 2008; pp. 1–38.
31. Hu, J.; Wu, W.; Zhong, C.; Liu, N.; Ouyang, C.; Yang, H.Y.; Yang, S.A. Three-dimensional honeycomb carbon: Junction line distortion and novel emergent fermions. *Carbon* **2019**, *141*, 417–426. [CrossRef]
32. Xie, L.; An, H.; He, C.; Qin, Q.; Peng, Q. Mechanical Properties of Vacancy Tuned Carbon Honeycomb. *Nanomaterials* **2019**, *9*, 156. [CrossRef]
33. Jiang, J.-W. Parametrization of Stillinger–Weber potential based on valence force field model: Application to single-layer MoS_2 and black phosphorus. *Nanotechnology* **2015**, *26*, 315706. [CrossRef]
34. Zhang, J.; Wang, C.; Adhikari, S. Fracture and buckling of piezoelectric nanowires subject to an electric field. *J. Appl. Phys.* **2013**, *114*, 174306. [CrossRef]
35. Wang, J.; Kulkarni, A.J.; Ke, F.J.; Bai, Y.L.; Zhou, M. Novel mechanical behavior of ZnO nanorods. *Comput. Methods Appl. Mech. Eng.* **2008**, *197*, 3182–3189. [CrossRef]

© 2019 by the authors. Licensee MDPI, Basel, Switzerland. This article is an open access article distributed under the terms and conditions of the Creative Commons Attribution (CC BY) license (http://creativecommons.org/licenses/by/4.0/).

MDPI
St. Alban-Anlage 66
4052 Basel
Switzerland
Tel. +41 61 683 77 34
Fax +41 61 302 89 18
www.mdpi.com

Nanomaterials Editorial Office
E-mail: nanomaterials@mdpi.com
www.mdpi.com/journal/nanomaterials

www.ingramcontent.com/pod-product-compliance
Ingram Content Group UK Ltd.
Pitfield, Milton Keynes, MK11 3LW, UK
UKHW060124010826
14089UKWH00006B/146

9 783039 286324